Pseudomonas aeruginosa - Biofilm Formation, Infections and Treatments

Edited by Theerthankar Das

Published in London, United Kingdom

IntechOpen

Supporting open minds since 2005

Pseudomonas aeruginosa - Biofilm Formation, Infections and Treatments
http://dx.doi.org/10.5772/intechopen.87468
Edited by Theerthankar Das

Contributors
Theerthankar Das, Silvia Labovská, Rahman Chelab, Gianmarco Mangiaterra, Nicholas Cedraro, Francesca Biavasco, Mehdi Amiri, Hussein O. M. Al-Dahmoshi, Noor Salman Al-Khafaji, Raad D. Al-Obaidi

First published in London, United Kingdom, 2021 by IntechOpen
IntechOpen is the global imprint of INTECHOPEN LIMITED, registered in England and Wales, registration number: 11086078, 5 Princes Gate Court, London, SW7 2QJ, United Kingdom
Printed in Croatia

British Library Cataloguing-in-Publication Data
A catalogue record for this book is available from the British Library

Additional hard and PDF copies can be obtained from orders@intechopen.com

Pseudomonas aeruginosa - Biofilm Formation, Infections and Treatments
Edited by Theerthankar Das
p. cm.
Print ISBN 978-1-83968-647-4
Online ISBN 978-1-83968-648-1
eBook (PDF) ISBN 978-1-83968-649-8

Meet the editor

Dr. Theerthankar Das (Department of Infectious Diseases and Immunology, School of Medical Sciences, University of Sydney, Australia) is an experienced microbiologist. He joined the University of Sydney after being awarded the prestigious University of Sydney Fellowship in 2015. His primary research focuses on the development of novel strategies to disrupt bacterial biofilms. In recent years, he has won various research funding/grants from Sydney University and the Australian Government valued at more than $4.5 million. To date, Dr. Das has authored/co-authored twenty-eight publications and five book chapters in eminent journals and books. He is also a reviewer for many high-impact scientific journals. Dr. Das currently supervises Ph.D. students and teaches first-year Medical and Advanced Medical Bacteriology students.

Contents

Preface XI

Chapter 1 1
Introductory Chapter: Understanding Infections Caused by Opportunistic Bacterial Pathogens
by Theerthankar Das

Chapter 2 7
Pseudomonas aeruginosa: Diseases, Biofilm and Antibiotic Resistance
by Hussein Al-Dahmoshi, Raad D. Al-Obaidi and Noor Al-Khafaji

Chapter 3 21
Pseudomonas aeruginosa as a Cause of Nosocomial Infections
by Silvia Labovská

Chapter 4 33
Pseudomonas aeruginosa Biofilm Lung Infection in Cystic Fibrosis: The Challenge of Persisters
by Gianmarco Mangiaterra, Mehdi Amiri, Nicholas Cedraro and Francesca Biavasco

Chapter 5 55
Pseudomonas aeruginosa Secreted Biomolecules and Their Diverse Functions in Biofilm Formation and Virulence
by Theerthankar Das

Chapter 6 75
Chemotherapy and Mechanisms of Action of Antimicrobial Agent
by Rahman Laibi Chelab

Preface

I am greatly thankful to IntechOpen for inviting me to serve as editor of this book on *Pseudomonas aeruginosa - Biofilm Formation, Infections and Treatments*.

This book covers a wide array of subjects relevant to bacterial biofilms specifically focusing on *P. aeruginosa* and associated infections. The principal objective of this book is to provide readers with a clear and comprehensive overview of biofilm formation and its detrimental impacts. In addition, this book also examines topics related to biosynthesis virulence factors by *P. aeruginosa* to facilitate biofilm formation, antibiotic resistance mechanisms, and infections. Biofilms and associated infections have a huge impact on human health, livestock, agriculture, and the world's economy overall. Thus, it is of paramount importance for scientists, medical professionals, healthcare workers, and the public to gain knowledge on this recurring issue and do every bit to sustain biofilms and their damaging impacts.

P. aeruginosa is an opportunistic Gram-negative bacterium that easily thrives and survives in various biotic and abiotic environments. Colonization and establishment of this bacterium species as biofilms, especially in hospitals and other communal environments, leads to contamination of water and food and is the leading cause of healthcare-associated or nosocomial infections. *P. aeruginosa* is also the most common pathogen in patients with cystic fibrosis, urinary tract infection, and wounds. Its antibiotic resistance is a major cause of mortality and morbidity. Understanding the mechanisms by which *P. aeruginosa* establishes as biofilms, evades antibiotics and antiseptics, and triggers life-threatening infections in humans is very important. Research on *P. aeruginosa* has been in the limelight for many decades as evidenced by the hundreds of journals, conferences, and projects in this field. Published research papers, conferences, and opinions from expert scientists, clinicians, and healthcare workers in recent years have undoubtedly enhanced the scientific basis for *P. aeruginosa*-associated infections and treatments.

To this end, I would also like to express my gratitude to all the scientists and researchers from different research institutes and universities who put immense effort into writing their chapters for this book.

Theerthankar Das
Department of Infectious Diseases and Immunology,
School of Medical Sciences,
The University of Sydney,
Sydney, Australia

Chapter 1

Introductory Chapter: Understanding Infections Caused by Opportunistic Bacterial Pathogens

Theerthankar Das

1. Introduction

Infectious diseases caused by virus, bacteria and fungi represent a major apprehension globally in terms of detrimental public health and economy. Some of the infectious agents such as virus (e.g., Coronavirus, influenza, Ebola, chicken-pox), bacteria (e.g., tuberculosis, cholera, whooping cough) are highly contagious and are responsible for communicable diseases. Communicable diseases spread from person to person through close contact including touching and kissing, also via coughing and sneezing, contamination of food and water. Many bacteria are also opportunistic pathogens and are commonly cause mild–moderate infections including sore throat, acne, tooth decay, urinary tract infections, cellulitis/skin infection, sexually transmitted infection, bacterial vaginosis, peptic/stomach ulcer, keratitis/eye infection, to severe/life-threatening infections such as pneumoniae, septicaemia/sepsis, meningitides in humans, animals, and birds. Most opportunistic bacteria exist as a commensal flora within the host body (gastrointestinal tract, skin, mucosal, oral, and nasal cavity, urogenital tract) and commonly found in abiotic surfaces (water, food, soil) in the environment [1, 2]. Under normal conditions i.e., in healthy people these bacterial pathogens do not cause infections. Infections caused by opportunistic bacteria are primarily triggered by either invasion of host commensal bacteria or bacteria from environmental sources gets into host bodily tissue [1, 2]. However, these opportunistic bacteria primarily target and cause fatal infections in immunocompromised people including acquired immunodeficiency syndrome (AIDS/HIV positive) patients, cancer patients (treated with immunosuppressive drugs, corticosteroids), hospital admitted patients for surgery, patients with underlying diseases such as cystic fibrosis, diabetes [2]. Most common examples of opportunistic bacteria found in mammals, birds and environment are *Pseudomonas aeruginosa*, *Staphylococcus aureus*, *Staphylococcus epidermidis*, *Escherichia coli*, *Klebsiella pneumoniae*, *Neisseria meningitides*, *Acinetobacter baumannii*, *Helicobacter pylori*, *Streptococcus pneumoniae*, *Streptococcus pyogenes*, *Streptococcus mutans*, *Clostridioides difficile*, *Legionella pneumophila*, *Propionibacterium acnes*, etc. On the other hand, mammalian and bird's body also host different species of good bacterial species (probiotics) which are essential for general wellbeing [3]. Lactobacillus and Bifidobacteria species are the common example of probiotics bacteria that are present in the mammalian digestive tract, they aid in maintaining daily healthy lifestyle includes food digestion, balancing pH of the body, alleviate symptoms of Gastroesophageal reflux diseases (GERD) such as heartburn, acid-reflux [4, 5].

IntechOpen

2. Biofilms associated infection and its impact on human health and economy

Infection is predominantly triggered by biofilm formation. It is projected that more than 60% of all bacterial infections are associated with biofilms [6]. In simple words, biofilms mean colonization of bacteria on biotic and abiotic surfaces. Biofilm is the most supreme stage of bacterial lifestyle. Biofilm formation involves various stages to instigate with initial adhesion (reversible adhesion) of individual bacteria (planktonic stage) through their cell appendages (flagella, fimbriae, pili, that are anchored to bacterial cell surface) to host bodily surfaces such as skin, mucosa, and teeth also on medical devices such as implants, catheters, contact lenses and on water pipes, sinks, bathtub. The second steps are microcolony formation and biosynthesize of numerous exogenous biopolymers by bacteria such as nucleic acids (extracellular DNA, RNA), proteins, polysaccharides, virulence factors and metabolites that aids them in irreversible/strong adhesion to the surfaces, cell-to-cell adhesion, and foundation for initial architecture for biofilm. Later stages include further maturation and growth of bacteria within the microcolony and production of biomolecules and formation of robust three-dimensional biofilm and finally dispersion of individual bacteria from the mature biofilms to initiate colonization at new site [7, 8]. This close vicinity of the bacterial cells within the biofilm empowers exchange and distribution of various essential products includes nutrients, genetic materials, proteins, metabolites and other small molecules for the fitness, growth, and survival of bacterial cells and exclusion of toxic end products [9]. The complex biofilm architecture defends the bacterial cells within them from antibiotics, antiseptics, soaps and detergents, physical shear forces and the host immune system [10].

The prevalence and persistence of biofilm associated infections has direct adverse impact on human health and World's economy including costing billions of dollars annually for treatment across several different sectors such as wound and burns treatment, dentistry, endocarditis, bronchitis's, cystic fibrosis and surgical (hip, knee joints, pacemakers, cardiac valves) and non-surgical (contact lenses, urinary catheters, artificial teeth) implants [11, 12]. For instance, United States of America spends approximately 94 billion USD a year with more than half a million deaths related to biofilms [13]. Biofilm related Hospital-acquired infections/nosocomial infection includes pneumoniae, surgical site infections, Urinary tract infections (UTIs), blood stream infections alone cost USA health care 11 billion USD with approximately 2 million cases per year and is responsible for fourth leading cause of deaths in USA [6, 14, 15]. In general, it is speculated that nosocomial infection in patients becomes apparent within 48 hrs of early patient care [14]. Australia's health system already expended $909 million annually for treatment associated with Urinary tract infections (UTIs). To note: the ratio of patients affected with UTIs in Australia is estimated to be 1 out of 2 women and 1 out of 20 men in their lifetime. UTIs associated implications in Australia also resulted in over 2.5 million visits to clinics and 75,000 hospital stay yearly [16]. As per statistics, report by European centre for disease prevention and control, nosocomial infection rate in European Union countries is soar, estimated to be around 3 million people get infected and around 50,000 death associated with it per year [17]. Primary factors that trigger the increase in infection rate, morbidity and mortality and associated treatment cost are due to poor hygiene, malnutrition, and lack of sanitation especially in the low-income countries, also misuse of antibiotic in food industry (agriculture, livestock, dairy) and unwarranted prescription (e.g., antibiotics prescribed to patients for common cold). Development of multidrug resistance bacteria or superbugs further escalates infection rate and associated treatment cost and death rate. News article published by leading newspaper "Times of India" reported mortality rate in India due to superbugs is 13% in comparison to 2–7% in developed countries [18]. Biofilms also posses' serious threat to food sector including agriculture, dairy, and livestock. It is estimated that infections in plants by microbial biofilms add to 10% of global food supply loss and directly contribute to foodborne infections [11]. Bovine mastitis, potentially fatal mammary gland infection/inflammation of the udder in cow, caused by bacteria attribute to loss of two billion dollars to the US dairy industry [11].

3. Pseudomonas aeruginosa a critical opportunistic bacterium

Pseudomonas aeruginosa, is one such opportunistic Gram-negative rod-shaped bacterial pathogen known for its ubiquity. World health organization (WHO) have placed *P. aeruginosa* in top priority (critical) organism list considering its intrinsic antibiotic resistance profile and remarkable ability to acquire tolerance to antibacterial agents [19]. In addition, *P. aeruginosa* forms robust biofilm and triggers severe infections especially in immunocompromised and hospital admitted patients. *P. aeruginosa* commonly found in human gastrointestinal tract, skin, soil, water, meat, plants, and vegetables and one of the leading causes for blood stream infection, UTI, microbial keratitis, wound and burn infection, HIV/AIDS patients, in ICU patients (ventilator associated pneumoniae) and a leading death cause in cystic fibrosis patients. *P. aeruginosa* associated hospital-acquired infections ranges between 10 and 15% globally [20]. Global epidemiology survey on *P. aeruginosa*, recorded numerous antibiotic resistance strains isolated from infected patients. These isolates are resistance to many antibiotics (carbapenem, gentamicin, ciprofloxacin, tobramycin, meropenem, and others) which are commonly used to treat infected patients [21–24]. This bacterium secretes numerous biomolecules such as DNA, proteins, polysaccharides, pyocyanin, rhamnolipids, siderophores which supports them in colonization at infection site and spread virulence in host and shield them from antibacterial agents [25]. In this book, we elaborated on general bacterial biofilm and in specifically focused on mechanism of *P. aeruginosa* biofilm formation, pathogenicity, antibiotic resistance, and treatment. The collections of chapters in this book will enlighten different end users including infectious diseases scientist, medical professional, medical and microbiology students and public.

Author details

Theerthankar Das
Department of Infectious Diseases and Immunology, School of Medical Sciences, The University of Sydney, Australia

*Address all correspondence to: das.ashishkumar@sydney.edu.au

References

[1] Ribet D, Cossart P. How bacterial pathogens colonize their hosts and invade deeper tissues. Microbes & Infect. 2015;17(3):173-183. Doi: https://doi.org/10.1016/j.micinf.2015.01.004

[2] Jose JR, Brown SJ. Opportunistic bacterial, viral and fungal infections of the lung. Medicine (Abingdon). 2016;44(6):378-383. doi: 10.1016/j.mpmed.2016.03.015

[3] [Internet]. Available from: http://www.melbournebirdvet.com/use-of-probiotics/.

[4] Aiba Y, Nakano Y, Koga Y, Takahashi K, Komatsu Y. A highly acid-resistant novel strain of *Lactobacillus johnsonii* No. 1088 has antibacterial activity, including that against *Helicobacter pylori*, and inhibits gastrin-mediated acid production in mice. Microbiologyopen. 2015;4: 465-474. doi: 10.1002/mbo3.252.

[5] Urita Y, Goto M, Watanabe T, Matsuzaki M, Gomi A, Kano M, Miyazaki K, Kaneko H. Continuous consumption of fermented milk containing *Bifidobacterium bifidum* YIT 10347 improves gastrointestinal and psychological symptoms in patients with functional gastrointestinal disorders. Biosci. Microb. Food H. 2015;34:37-44. doi: 10.12938/bmfh.2014-017.

[6] [Internet]. Available from: https://lbtinnovations.com/projects/biofilm/

[7] Berne C, Ellison CK, Ducret A, Brun YV. Bacterial adhesion at the single-cell level. Nat. Rev. Microbiol. 2018;16, 616-527. doi: https://doi.org/10.1038/s41579-018-0057-5

[8] Hollmann B, Perkins M, Walsh D. Biofilms and their role in pathogenesis. [Internet]. Available from: https://www.immunology.org/public-information/bitesized-immunology/pathogens-and-disease/biofilms-and-their-role-in

[9] Nazzaro F, Fratianni F, D'Acierno A, Coppola R, Jesus F, Gomez da Cruz AA, Feo VD. Essential Oils and Microbial Communication. IntechOpen, 2019.

[10] Saxena P. Joshi Y, Rawat K, Bisht R. Biofilms: Architecture, Resistance, Quorum Sensing and Control Mechanisms. Indian J Microbiol. 2019; 59(1):3-12.doi: 10.1007/s12088-018-0757-6.

[11] [Internet]. Available from: https://www.paconsulting.com/insights/rise-to-the-biofilm-challenge/

[12] Rao AN, Avula MN, Grainger DW. Aging and the Host Response to Implanted Biomaterials. Book: Host response to Biomaterials. Academic Press. 2015.

[13] Wolcott RD, Rhoads DD, Bennett ME, Wolcott BM, Gogokhia L, Costerton JW, Dowd SE. Chronic wounds and the medical biofilm paradigm. J Wound Care. 2010;19(2): 45-46, 48-50, 52-43.

[14] Wenzel RP. Health Care–Associated Infections: Major Issues in the Early Years of the 21st Century. Clin Infect Dis. 2007; 15;45. doi: 10.1086/518136.

[15] [Internet]. Available from: https://cleanhands-safehands.com/hais-4th-leading-cause-of-death/

[16] [Internet]. Available from: https://outbreakproject.com.au/2020/11/17/australias-multi-billion-dollar-superbug-crisis/

[17] The First European Communicable Disease Epidemiological Report. 2007. European Centre for Disease Prevention and Control https://www.ecdc.europa.eu/sites/portal/files/media/en/

publications/Publications/0706_SUR_First_%20Annual_Epidemiological_Report_2007.pdf

[18] [Internet]. Available from: https://timesofindia.indiatimes.com/india/superbugs-kill-more-in-india-than-globally-mortality-rate-is-13/articleshow/66674259.cms

[19] [Internet]. Available from: https://www.who.int/news/item/27-02-2017-who-publishes-list-of-bacteria-for-which-new-antibiotics-are-urgently-needed

[20] Blanc DS, Petignat C, Janin B, Bille J, Francioli P. Frequency and molecular diversity of *Pseudomonas aeruginosa* upon admission and during hospitalization: a prospective epidemiologic study. Clin Microbiol Infect. 1998;4:242-7. doi.org/10.1111/j.1469-0691.1998.tb00051.x.

[21] Sievert DM, Ricks P, Edwards JR, Schneider A, Patel J, Srinivasan A, Kallen A, Brandi Limbago B, Fridkin S, Antimicrobial-resistant pathogens associated with healthcare-associated infections: summary of data reported to the National Healthcare Safety Network at the Centers for Disease Control and Prevention, 2009-2010. Infect Control Hosp Epidemiol. 2013. 34(1):1-1)14. Doi: 10.1086/668770.

[22] Kumari M, Khurana S, Bhardwaj N, Malhotra R, Mathur P. Pathogen burden & associated antibiogram of Pseudo monas spp. in a tertiary care hospital of India. Indian J Med Res. 2019 Feb;149(2):295-298. Doi: 10.4103/ijmr.IJMR_14_18.

[23] Khan MA, Faiz A. Antimicrobial resistance patterns of Pseudomonas aeruginosa in tertiary care hospitals of Makkah and Jeddah.Ann Saudi Med. 2016;36(1):23-28. Doi: 10.5144/0256-4947.2016.23.

[24] Rashid A, Chowdhury A, Rahman SHZ, Begum SA, Muazzam N. Infections by Pseudomonas aeruginosa and antibiotic resistance pattern of the isolates from Dhaka Medical College hospital. Bangladesh J Med Microbiol. 2007;1:48-51.

[25] Das T, Manoharan A, Glasbey T, Whiteley G, Jim Manos. *Pseudomonas aeruginosa* biofilms and infections: Roles of extracellular molecules. Book Name: New and Future Developments in Microbial Biotechnology and Bioengineering: Microbial Biofilms. Elsevier; 2019.

Chapter 2

Pseudomonas aeruginosa: Diseases, Biofilm and Antibiotic Resistance

Hussein Al-Dahmoshi, Raad D. Al-Obaidi and Noor Al-Khafaji

Abstract

Pseudomonas aeruginosa is Gram negative bacteria that can adapt to extreme environmental conditions and withstand to different antibacterial agents. It si responsible for arrays of infections both community and hospital acquired especially ICU infections. Respiratory tract infection, blood stream infection, wound infection, burn infection, and urinary tract infections ware top five *P. aeruginosa* infections. Additionally as an opportunistic bacteria, it may be associated with healthcare infections in intensive care units (ICUs), ventilator-associated pneumonia (VAP), central line-associated blood stream infections, surgical site infections, otitis media, and keratitis. *P. aeruginosa* can form biofilms as self-produced extracellular matrix to protects the cells from antibiotics and the host immune response. Antibiotic resistance was an prominent feature of this pathogen and can donate it one of the three resistance patterns: Multidrug (MDR), extensive drug (XDR) and pan drug resistance. It exploit many resistance mechanisms ranged from overexpression of drug efflux systems protein, modifying enzyme production, reducing the permeability and using shelters like biofilms.

Keywords: *P. aeruginosa*, UTIs, RTIs, wound infections, antibiotic resistance, biofilm

1. Introduction

Pseudomonas aeruginosa (P. aeruginosa) is considered as a part related to normal intestinal flora as well as a considerable pathogen that is accountable for various ICU-acquired infections in patients who are critically ill. The nosocomial infections related to such organism involve meningitis, blood stream infections, urinary tract infections, respiratory tract infections, wound infections and otitis media [1, 2]. Study samples have been gathered from patients with the next disease: *P. aeruginosa* ear infection might be categorized into otitis media, malignant outer otitis, simple external otitis (or swimmer ear) and perichondritis. Particularly, otitis media can be defined as one of the middle ear inflammations impacting the pediatric population and might be divided as chronic and acute [3]. *P. aeruginosa* can be specified as one of the opportunistic bacteria related to health-care infections in VAP, ICU, central line related blood stream infections, surgical site infections, urinary tract infections, burnt wounds, keratitis and otitis media [4, 5]. *P. aeruginosa* might be specified as a bacterial pathogen that causes extreme chronic infections in the immuno-compromised individuals. The capability of *P. aeruginosa* for creating biofilm, that were communities regarding the cells which are encased in self produced extracellular matrix, protecting the cells from antibiotics as well as host immune responses [6]. The biofilm have the ability of augmenting the persistent infections related to *P. aeruginosa* and high antibiotic resistance level in comparison to planktonic bacterial cells, whereas the treatment of infections resulting from such organism are difficult due to the existence of its innate resistance to various antibiotics (β-lactam and penem group of antibiotics), along with its capability for acquiring more resistance mechanism to several antibiotic types, such as fluoroquinolones, aminoglycosides and Beta-lactams [7]. *P. aeruginosa* implicated disease are indicated in the following way:

1.1 *P. aeruginosa* associated respiratory tract infection

P. aeruginosa can be defined as a significant organism resulting in chronic infections in the bronchiectasis, due to its ability for maintaining virulence in spite of antibiotic therapies via creating biofilm and developing antimicrobial resistance. In addition, bronchiectasis is one of the chronic airway diseases specified via irreversibly damaged as well as dilated bronchi resulting in recurrent bronchial sepsis episodes. This leads to poor mucus clearance, and vicious cycle related to persistent bacterial colonization, inflammation, airway obstruction along with progressive destruction of the tissues [8]. Also, *P. aeruginosa* in the cystic fibrosis (CF) bronchiectasis was related to steeper reduction in the function of lungs and elevated mortality and morbidity. Its existence in bronchiectasis is related to diseases with more severity, yet if there was direct impact on the progression of disease or if *P. aeruginosa* one of the indicators of current clinical severity still debated [9, 10]. Also, the rate of *P. aeruginosa* chronic infections in patients experiencing bronchiectasis vary from 9 and 31%, and its commonness in large, multi-centre population from many nations yet to be evaluated [11, 12]. Furthermore, CF lung was hostile, heterogeneous and stressful environments for the invading bacteria, while the populations of *P. aeruginosa* should have the ability of overcoming such issues for persisting and surviving, whereas the postulated stressors in CF lung involve osmotic stress because of mucus, oxidative and nitrosative stresses because of host responses, sub-lethal antibiotics' concentrations, along with the existence of other microorganisms [13].

1.2 *P. aeruginosa* associated wound and burn infection

P. aeruginosa is specified as a major pathogen isolated from the burn patients worldwide [14], also it is opportunistic bacterium related to VAP, ICU, burns and surgical site infections [15] and it is a significant pathogens included in burn infections [16] and one of the main nosocomial pathogens in the burn patients, and quickly acquiring antibiotic resistance; therefore, develop efficient therapeutic method was one of the main strategies to combat the infections [17]. Particularly in burn centers, the progressive increase and high occurrence of MDR *P. aeruginosa* threating patients with extreme burn injuries [18, 19], while burn wound infections were major complications happening following the burn injuries and might be related to dangerous clinical complications and elevated mortality and morbidity [20]. In addition, burn injuries includes the primary host's barrier, the skin, that is directly placing hosts at risks of infections [21], whereas the burn wounds were main public-health problems worldwide. Infections are difficult problems in burn patients, since the skin, one of the barriers against microbes, was destroyed and immunity agents have no ability for reaching the infection sites. There were correlations between the infection severity and the burn's extent [22]. *P. aeruginosa* is a typical bacterium in the nosocomial infections, particularly in burn units.

Furthermore, burn patients, due to the loss of skin barrier, showed high vulnerability to infections [23]. Novel therapeutic agents against the *P. aeruginosa*, improve the effectiveness of present antimicrobial agents and degrade biofilm in burn wounds, were needed [24], such bacterium is causing 75% of deaths in the burned patients, as it might be developing persistent biofilm related to infections, expressing many virulence factors, as well as mechanisms of antibiotic resistance. A few of such virulence factors have been proteases like elastase and alkaline protease, or toxic metabolites including pyocyanin and microorganisms with the ability of producing cyanide, that is inhibiting the cytochrome oxidase regarding host cells [25]. Furthermore, multiple antibiotic resistant *P. aeruginosa* was a considerable cause of burn wound infections and, soft tissue and skin infections. Due to its resistance to majorly utilized antiseptics and antibiotics, there was lack of therapeutic options for efficient treatments [26]. Usually, *P. aeruginosa* attacking patients with wound and burn infections, in which more complications of primary condition, might happen and often causing bacteremia [27].

1.3 *P. aeruginosa* associated urogenital infection

UTC is one of the major microbial diseases with considerable economic impacts on society [28]. Even though that almost all UTIs were resulting from *E. coli*, a lot of epidemiological reports showed an increase in the infection's rates resulting via a few of opportunistic organisms involving *P. aeruginosa*. Also, the pathogenesis mechanisms and antimicrobial susceptibility regarding *P. aeruginosa* were badly understood [29]. UTC is a major microbial infection in humans and representing considerable burden on health-care systems. UTI might be simple, in terms of affecting healthy people, or complicated, in terms of impacting people with compromised host defenses and/or urodynamics, like the ones with urinary catheter, while there were some differences between un-complicated UTI as well as catheter associated UTIs (CAUTIs) in the clinical manifestations, pathophysiology and causative organisms. Thus, uncomplicated CAUTI and UTI might not be similarly approached, or the risks of complications and treatment failure might be increased [30]. Also, complicated UTI (cUTI), occurs in immuno-compromised patients or in the patients with functional or structural abnormalities of the urinary tract (UT), were related to high treatment failure rates and dangerous complications, particularly relapse as well as the development related to antibiotic resistance [31]. cUTIs is from the major healthcare-related infections.

In patients experiencing cUTI, *P. aeruginosa* is deserving distinctive attention, due to the fact that it might be affecting patients experiencing considerable underlying conditions [32]. Also, *P. aeruginosa* is a main nosocomial uropathogen. In addition, *P. aeruginosa* is considered as the 3rd major pathogen resulting in hospital-acquired CAUTI [33]. It might be tolerating a lot of physical conditions and a lot of antibiotics via various mechanisms of resistance [34]. The continuous increase in antibiotic resistance all over the world is disturbing [28]. Increasing the multi-drug resistance in the bacterial uropathogens is emerging and considerable public-health problems [35]. Catheter-associated UTC (CAUTI) is responsible for 40% of the nosocomial infections in hospitalized patients [33]. *P. aeruginosa* a multifaceted pathogens resulting in many biofilm mediated infections, involving CAUTIs. The majority of catheter-associated infections were caused by the own perineal flora of patient, yet the existence of catheter is increasing the chances of being colonized through cross transmission regarding nosocomial bacteria too.

The majority of episodes related to short term catheter associated bacteriuria were asymptomatic and resulting from single organisms, whereas long term catheterization is promoting colonization and multi-bacterial infections. The prolonged duration related to catheterization bacteriuria was specified universal due to the biofilm' formation on the surface of catheter [36]. In addition, the high occurrence of CAUTI in the hospitals, their clinical manifestations, like cystitis, urethritis, meningitis, pyelonephritis, death and urosepsis, also the related economic challenges underscoring the requirement for the infections' management, while *P. aeruginosa* might be resulting in complicated UTIs, especially in individuals with catheters, that might result in pyelonephritis, whereas a few sub-groups appearing with more susceptibility to infections, like women and elderly, the contributions of other host factors in addition to the bacterial virulence factors for successful infections still fairly understudied. Furthermore, *P. aeruginosa* UTIs were extremely antibiotic resistant and requiring intensive and costly treatment [37].

1.4 *P. aeruginosa* associated otitis media

Otitis media (OM) can be considered as a major cause of fever as presentation in pediatric population. Chronic suppurative otitis media (CSOM) (also termed as chronic otitis media) is one of the ear diseases stages where there are ongoing chronic infections related to middle ear with no intact tympanic membrane, such disease is one of the inflammations of middle ear as well as mastoid cavity, while the characteristic presentation was persistent or chronic otorrhoea throughout (2–6) weeks via perforated tympanic membrane [38]. In addition, OM considered as group of the complex inflammatory disorders impacting the middle ear that might be chronic or acute [39]. CSOM (also referred to as chronic otomastoiditis, chronic active mucosal otitis media and chronic tympanomastoiditis), is one of the inflammations of middle ear as well as mastoid cavity, present with recurrent ear discharge or otorrhea via perforated tympanic membrane [40]. Chronic (OM) is a perforation related to tympanic membrane with infection. Mostly, it occurs in underdeveloped nations.

Culture and sensitivity reports are showing the main pathogens accountable for chronic OM were *P. aeruginosa* and *S. aureus* [41]. Bacteria's dispersal from biofilm in the middle ear, serve as bacterial reservoir, might be explaining the recurrent as well as chronic nature related to CSOM [42]. *P. aeruginosa* is one of the significant CSOM pathogens showing multiple resistances to the antibiotics with increase in frequency and make the treatment of patients extra difficult [43]. *P. aeruginosa* is a major organism result in CSOM, is one of the notorious pathogens with MDR attribute [44]. Also, *P. aeruginosa* is invading the human middle ear epithelial cells (HMEECs) as well as inducing cytoskeletal rearrangements [39], while the antibiotics resistance that is considered as worldwide health challenge is not a future threat anymore, yet current problem to all clinical setting's facets. Therefore, treating OM effusion is a main concern [45].

1.5 *P. aeruginosa* associated meningitis

Community acquired meningitis resulting from *P. aeruginosa* has extremely increased mortality rates and uncommon [46]. It is related to prior neurosurgical procedure and hospital-related onset. Generally, bacterial meningitis is developing many cerebrovascular complications, from which the intracerebral hemorrhage is rare [47]. In addition, Multidrug Resistant (MDR) strains were identified in patients with neurosurgical interventions and patient with nosocomial exposure. This study is providing one of the fatal cases regarding community-acquired MDR pseudomonal meningitis [48]. One of the uncommon causes of ventriculitis and meningitis is *P. aeruginosa*, yet is commonly related to considerable mortality and morbidity [49]. Majorly, there is history of neurosurgical procedure in patients who develop *P. aeruginosa* meningitis [50]. Also, *P. aeruginosa* neurosurgical meningitis is one of the uncommon entities typically associated to elevated rates or mortality and with intraventricular catheters [51]. Adult bacterial meningitis (ABM) resulted from *P. aeruginosa* was typically because of nosocomial infections and typically identified in patients experiencing post neurosurgical state [49]. Each year, approximately 16,000 are dying due to ABM [52]. Meningitis resulting from extensively drug resistant (XDR) or multidrug-resistant (MDR). Gram negative bacillary meningitis (GNBM) resulting in considerable limitations in present options of treatment [53]. One of the main challenges in antibiotic selection is the increase in meningitis resulting from extreme drug resistant bacillary [52], while antibiotic resistance in multiple *P. aeruginosa* strains is quickly-developing clinical problem [54]. The options of treatment were further limited with the involvement of central nervous system (CNS), since the colistin based regimens were disadvantaged via poor blood brain barrier penetration, frequently related to inadequate microbiological and clinical success. There was a recent increase in using intrathecal colistin and it is one of the alternatives to manage infections related to central nervous system resulting from MDR bacteria [46]. whereas *P. aeruginosa* ventriculitis and meningitis were mainly nosocomial and associated to prior neurosurgery. There is high difficulty in diagnosing as Cerebrospinal fluid Gram film as well as meningism were insensitive markers [49].

2. Biofilm formation and antibiotic resistance

Extracellular matrix is vital feature related to biofilm communities, it is surrounding the resident bacteria and it includes matrix proteins, lipid vesicles, exopolysaccharides and extracellular DNA (eDNA), whereas the 3 exopolysaccharides regarding *P. aeruginosa* biofilm matrix (alginate, Pel and Psl) [55]. Mainly, the biofilm includes bacterial derived exopolysaccharides which is protecting encapsulated bacteria from host immune cells as well as antibiotics [56]. In addition, biofilm are considered to be widespread in their nature and constituting a significant strategy carried out via microorganisms for surviving in often harsh conditions of environment. They might be effectives or leaving bad effect especially when created on medical devices or in industrial settings. Thus, studying the elimination and formation of biofilm is significant for a lot of discipline [57]. The ability of *P. aeruginosa* for creating biofilm, that were cells' communities which are encased in self produced extracellular matrix, protecting the cells from antibiotics as well as host immune [58]. Also, *P. aeruginosa* is considered opportunistic, nosocomial bacterial pathogen forming persistent infections because of creating protective communities, referred to as biofilm. Furthermore, biofilm is a significant virulence factor in *P. aeruginosa* and has considerable roles in antibiotic resistance as well as chronic burn wound infections [24], while the biofilm of *P. aeruginosa* are contributing to its survival on the abiotic and biotic surfaces and representing main clinical threat because of their increased tolerance to the antibiotics [59]. As soon as forming the biofilm, the bacteria embedded in it were recalcitrant to the anti-microbial treatment along with host immune defenses [60].

Biofilm have been specified as complex microbial communities which contain micro-colonies and surrounded by self-created extracellular polysaccharide matrix, while the biofilm matrix in *P. aeruginosa* includes 3 different exopolysaccharides: Pel, Psl and alginate. Also, the alginate can be defined as one of the polymers which contain α-L-guluronic acid and β-D-mannuronic acid with considerable roles in the structural stability as well as biofilm protection, while Psl has been specified as a polysaccharide includes repeating pentasaccharide, containing L-rhamnose,

D-glucose and D-mannose. Psl was significant in the start of the formation of the biofilm and biofilm structure protection. Pel is specified as the 3rd polysaccharide that exists in *P. aeruginosa* biofilm and was glucose rich [61]. Furthermore, the biofilm cells showing increased resistance to the environmental pressures like anti-microbial agents compared to their planktonic form [62]. Also, its populations undergoing characteristic evolutionary adaptation throughout chronic infection related to CF lung, involving decreased virulence factors' production, transition to biofilm related lifestyles, and the evolution regarding high level antibiotics resistance, whereas the populations of *P. aeruginosa* in the chronic CF lung infections generally showing increased phenotypic diversity, involving clinically significant characteristics like antibiotics resistance and toxin production, and such diversity was dynamic throughout the time, which will make precise treatment and diagnosis challenging [63].

3. Antibiotic resistance

P. aeruginosa has been considered as a major cause related to nosocomial infections, also it is accountable for about 10% of all the hospital acquired infections in the world. It is still considered as one of the therapeutic challenges due to the high rates of mortality and morbidity related to it and the potential to develop drug resistance throughout the therapy. Also, standard antibiotic regimes against the *P. aeruginosa* were more and more unsuccessful due to the increase in drug resistance [64]. In addition, antibiotics resistance in the multiple strains related to *P. aeruginosa* was a clinical issue that is developing rapidly, while the definitions regarding multidrug resistance *P. aeruginosa* (MDRPA) was isolates resistant to minimum of 3 drugs from various antimicrobial categories, involving cephalosporinsand quinolones, aminoglycosides, carbapenems and anti pseudomonal penicillin were categorized as multidrug resistant. The development of antibiotic resistant bacteria in health-care is dangerous. With regard to health-care premises exactly ICUs were main microbial diversity sources. Recently, a few studies indicated that not just microbial diversity, yet also the drug resistant microbes majorly habitat in the ICUs.

Infections resulting from such organism were complicated to treat due to the existence of its innate resistance to various antibiotic types (Beta-lactam and penem group of antibiotics) as well as its capability for acquiring more resistance mechanism for a number of antibiotics classes, involving aminoglycosides, β-lactams and fluoroquinolones. With regard to molecular evolution microbes adopting many mechanisms for maintaining genomic plasticity [2], MDR isolates have been majorly specified via slow growth, cytotoxic type-III secretion system genotype, excellent biofilm forming capability, and the existence of more aminoglycoside modifying enzyme (AME) genes, non MDR isolates are re-sensitized following the inhibition regarding active efflux or improvement of membrane permeability, such target gene alteration along with the enzymatic drug modification that has been specified as the main quinolone mechanisms and aminoglycoside resistance in *P. aeruginosa* keratitis isolates [65]. Extensively drug-resistant *P. aeruginosa* (XDR-PA) that has been characterized as the strains remaining susceptible to only 1 or 2 anti-pseudomonal agent classes, became a serious issue because of a lack of effective anti-microbial treatment [66].

P. aeruginosa became resistant to a number of the antibiotics classes, which include the carbapenems, which have been viewed as reliable antibiotics for treating the multi-drug-resistant *P. aeruginosa* serious infections and have been viewed as a last-resort antibiotic therapy of the infections that have been caused by the carbapenem-resistant *Pseudomonas aeruginosa* has become more problematic, particularly with increasing the carbapenem resistance. Carbapenem was commonly utilized for the directed or empirical treatments when a PA infection has been suspected as a result of its natural resistance towards numerous antibiotic types [67]. None-the-less, the recent data from National Antimicrobial Resistance Surveillance, Thailand (NARST), has shown an increasing CRPA trend, from about 15% of infections in the period 2000–2005 to 30% in the period 2009–2013. The rate of the CRPA which is related healthcare-associated infection (HAI) increased in past year all over the world [68].

Pseudomonas aeruginosa isolates have intrinsic resistance to the majority of the antimicrobials through the chromosomal AmpC cephalosporinases as well as low permeability to the antimicrobials, and can be accumulating extra resistance determinants through acquiring elements of the mobile genetics. *Pseudomonas aeruginosa* is of a large genome (i.e. higher than 6 MB), a high proportion of the regulatory genes and a set of the virulence determinants. The capability of using several mechanisms, which includes a decrease in the external membrane permeability, produces antibiotic degradative enzymes, efflux pump expression, production of the alginate and resistance gene transfer, the bacteria enabled showing a high resistance level to most of the utilized antibiotics [69]. Several recent researches have reported alternative and complementary options of the treatment to the infections of the combat *Pseudomonas aeruginosa*. Quorum sensing inhibitors, probiotics, phages, vaccine antigens, antimicrobial peptides, and anti-microbial nano-particles have the possibility of acting against the drug resistant strains [64].

4. Conclusion

The current review conclude the implication of *P. aeruginosa* in arrays of diseases especially RTIs, UTIs and wound infections. The widespread of it may be due to their adaptation to different environmental conditions along with virulence traits especially biofilm formation and intrinsic and acquired antibiotic resistance strategies.

Conflict of interest

There is no 'conflict of interest' for this work.

Author details

Hussein Al-Dahmoshi*, Raad D. Al-Obaidi and Noor Al-Khafaji
Biology Department, College of Science, University of Babylon, Hilla, Iraq

*Address all correspondence to: dr.dahmoshi83@gmail.com

References

[1] Streeter, K. and Katouli, M., 2016. *Pseudomonas aeruginosa*: A review of their Pathogenesis and Prevalence in Clinical Settings and the Environment. INFECTION EPIDEMIOLOGY AND MICROBIOLOGY.2(1): 25-32. DOI:10.7508/iem.2016.01.008.

[2] Pachori, P., Gothalwal, R. and Gandhi, P., 2019. Emergence of antibiotic resistance *Pseudomonas aeruginosa* in intensive care unit; a critical review. Genes & diseases, 6(2), p.109.DOI: 10.1016/j.gendis.2019.04.001.

[3] Lieberthal, A.S., Carroll, A.E., Chonmaitree, T., Ganiats, T.G., Hoberman, A., Jackson, M.A., Joffe, M.D., Miller, D.T., Rosenfeld, R.M., Sevilla, X.D. and Schwartz, R.H., 2013. The diagnosis and management of acute otitis media. Pediatrics, 131(3), pp.e964-e999.DOI10.1542/peds.2012-3488

[4] Cohen, R., Babushkin, F., Cohen, S., Afraimov, M., Shapiro, M., Uda, M., Khabra, E., Adler, A., Ami, R.B. and Paikin, S., 2017. A prospective survey of *Pseudomonas aeruginosa* colonization and infection in the intensive care unit. Antimicrobial Resistance & Infection Control, 6(1), p.7. DOI 10.1186/s13756-016-0167-7

[5] Fournier, A., Voirol, P., Krähenbühl, M., Bonnemain, C.L., Fournier, C., Pantet, O., Pagani, J.L., Revelly, J.P., Dupuis-Lozeron, E., Sadeghipour, F. and Pannatier, A., 2016. Antibiotic consumption to detect epidemics of *Pseudomonas aeruginosa* in a burn centre: A paradigm shift in the epidemiological surveillance of *Pseudomonas aeruginosa* nosocomial infections. Burns, 42(3), pp.564-570.DOI: 10.1016/j.burns.2015.10.030

[6] Cassin, E.K. and Tseng, B.S., 2019. Pushing beyond the envelope: the potential roles of OprF in *Pseudomonas aeruginosa* biofilm formation and pathogenicity. DOI: 10.1128/JB.00050-19

[7] Schaible, B., Crifo, B., Schaffer, K. and Taylor, C.T., 2020. The putative bacterial oxygen sensor *Pseudomonas* prolyl hydroxylase (PPHD) suppresses antibiotic resistance and pathogenicity in *Pseudomonas aeruginosa*. Journal of Biological Chemistry, 295(5), pp.1195-1201..DOI: 10.1074/jbc.RA119.010033

[8] Pasteur, M.C., Bilton, D. and Hill, A.T., 2010. British Thoracic Society guideline for non-CFbronchiectasis. Thorax, 65(Suppl 1), pp.i1-i58.DOI: 10.1136/thx.2010.136119.

[9] Cullen, L. and McClean, S., 2015. Bacterial adaptation during chronic respiratory infections. Pathogens, 4(1), pp.66-89.DOI: 10.3390/pathogens4010066.

[10] Aliberti, S., Lonni, S., Dore, S., McDonnell, M.J., Goeminne, P.C., Dimakou, K., Fardon, T.C., Rutherford, R., Pesci, A., Restrepo, M.I. and Sotgiu, G., 2016. Clinical phenotypes in adult patients with bronchiectasis. European Respiratory Journal, 47(4), pp.1113-1122.DOI: 10.1183/13993003.01899-2015.

[11] McDonnell, M.J., Jary, H.R., Perry, A., MacFarlane, J.G., Hester, K.L., Small, T., Molyneux, C., Perry, J.D., Walton, K.E. and De Soyza, A., 2015. Non cystic fibrosis bronchiectasis: a longitudinal retrospective observational cohort study of *Pseudomonas* persistence and resistance. Respiratory medicine, 109(6), pp.716-726.DOI: 10.1016/j.rmed.2014.07.021.

[12] Suarez-Cuartin, G., Smith, A., Abo-Leyah, H., Rodrigo-Troyano, A., Perea, L., Vidal, S., Plaza, V., Fardon, T.C., Sibila, O. and Chalmers, J.D., 2017. Anti-*Pseudomonas aeruginosa*

IgG antibodies and chronic airway infection in bronchiectasis. Respiratory medicine, 128, pp.1-6.DOI: 10.1016/j.rmed.2017.05.001.

[13] Lopes, S.P., Azevedo, N.F. and Pereira, M.O., 2015. Microbiome in cystic fibrosis: shaping polymicrobial interactions for advances in antibiotic therapy. Critical reviews in microbiology, 41(3), pp.353-365.DOI: 10.3109/1040841X.2013.847898.

[14] Sousa, D., Ceniceros, A., Galeiras, R., Pértega-Díaz, S., Gutiérrez-Urbón, J.M., Rodríguez-Mayo, M., López-Suso, E., Mourelo-Fariña, M. and Llinares, P., 2018. Microbiology in burns patients with blood stream infections: trends over time and during the course of hospitalization . Infectious Diseases, 50(4), pp.289-296. DOI: 10.1080/23744235.2017.1397738

[15] López-Jácome, L.E., Garza Ramos-Martínez, G., Hernández-Durán, M., Franco-Cendejas, R., Romero-Martínez, D., Thi Nguyen, P.D., Maeda, T., González-Pedrajo, B., Díaz-Guerrero, M., Loarca, D. and Sanchéz Reyes, J.L., 2019. AiiM lactonase strongly reduces quorum sensing controlled virulence factors in clinical strains of *Pseudomonas aeruginosa* isolated from burned patients. Frontiers in microbiology, 10, p.2657.DOI: 10.3389/fmicb.2019.02657

[16] Rafla, K. and Tredget, E.E., 2011. Infection control in the burn unit. Burns, 37(1), pp.5-15.DOI: 10.1016/j.burns.2009.06.198

[17] Ranjbar, M., Behrouz, B., Norouzi, F. and Gargari, S.L.M., 2019. Anti-PcrV IgY antibodies protect against *Pseudomonas aeruginosa* infection in both acute pneumonia and burn wound models. Molecular immunology, 116, pp.98-105.DOI: 10.1016/j.molimm.2019.10.005

[18] de Almeida, K.D.C.F., Calomino, M.A., Deutsch, G., de Castilho, S.R., de Paula, G.R., Esper, L.M.R. and Teixeira, L.A., 2017. Molecular characterization of multidrug-resistant (MDR) *Pseudomonas aeruginosa* isolated in a burn center. Burns, 43(1), pp.137-143. DOI: 10.1016/j.burns.2016.07.002

[19] Dou, Y., Huan, J., Guo, F., Zhou, Z. and Shi, Y., 2017. *Pseudomonas aeruginosa* prevalence, antibiotic resistance and antimicrobial use in Chinese burn wards from 2007 to 2014. Journal of International Medical Research, 45(3), pp.1124-1137.DOI: 10.1177/0300060517703573

[20] Turner, K.H., Everett, J., Trivedi, U., Rumbaugh, K.P. and Whiteley, M., 2014. Requirements for *Pseudomonas aeruginosa* acute burn and chronic surgical wound infection. PLoS genetics, 10(7).DOI:

[21] Lopez, O.N., Cambiaso-Daniel, J., Branski, L.K., Norbury, W.B. and Herndon, D.N., 2017. Predicting and managing sepsis in burn patients: current perspectives. Therapeutics and clinical risk management, 13, p.1107 DOI: 10.2147/TCRM.S119938

[22] Anvarinejad, M., Japoni, A., Rafaatpour, N., Mardaneh, J., Abbasi, P., Shahidi, M.A., Dehyadegari, M.A. and Alipour, E., 2014. Burn patients infected with metallo-beta-lactamase-producing *Pseudomonas aeruginosa*: Multidrug-resistant strains. Archives of trauma research, 3(2).DOI: 10.5812/atr.18182

[23] Moghoofei, M., Fazeli, H., Poursina, F., Esfahani, B.N., Moghim, S., Vaez, H., Hadifar, S. and Safaei, H.G., 2015. Morphological and bactericidal effects of amikacin, meropenem and imipenem on *Pseudomonas aeruginosa*. Jundishapur journal of microbiology, 8(11)DOI: 10.5812/jjm.25250.

[24] Banar, M., Emaneini, M., Satarzadeh, M., Abdellahi, N., Beigverdi, R., van Leeuwen, W.B.

and Jabalameli, F., 2016. Evaluation of mannosidase and trypsin enzymes effects on biofilm production of *Pseudomonas aeruginosa* isolated from burn wound infections. PloS one, 11(10). DOI: 10.1371/journal.pone.0164622

[25] López-Jácome, L.E., Garza Ramos-Martínez, G., Hernández-Durán, M., Franco-Cendejas, R., Romero-Martínez, D., Thi Nguyen, P.D., Maeda, T., González-Pedrajo, B., Díaz-Guerrero, M., Loarca, D. and Sanchéz Reyes, J.L., 2019. AiiM lactonase strongly reduces quorum sensing controlled virulence factors in clinical strains of *Pseudomonas aeruginosa* isolated from burned patients. Frontiers in microbiology, 10, p.2657.DOI: 10.3389/fmicb.2019.02657

[26] Nagoba, B., Davane, M., Gandhi, R., Wadher, B., Suryawanshi, N. and Selkar, S., 2017. Treatment of skin and soft tissue infections caused by *Pseudomonas aeruginosa*—A review of our experiences with citric acid over the past 20 years. Wound Medicine, 19, pp.5-9.DOI: 10.1016/j.wndm.2017.09.005

[27] Inacio, H. S. M. et al. (2014) 'Phenotypic and genotypic diversity of multidrugresistant *Pseudomonas aeruginosa* isolates from bloodstream infections recovered in the Hospitals of Belo Horizonte, Brazil', Chemotherapy. Karger Publishers, 60(1), pp. 54-62.

[28] Ujmajuridze, A., Chanishvili, N., Goderdzishvili, M., Leitner, L., Mehnert, U., Chkhotua, A., Kessler, T.M. and Sybesma, W., 2018. Adapted bacteriophages for treating urinary tract infections. Frontiers in microbiology, 9, p.1832.DOI: 10.3389/fmicb.2018.01832

[29] Badamchi, A., Masoumi, H., Javadinia, S., Asgarian, R. and Tabatabaee, A., 2017. Molecular detection of six virulence genes in *Pseudomonas aeruginosa* isolates detected in children with urinary tract infection. Microbial pathogenesis, 107, pp.44-47. DOI: 10.1016/j.micpath.2017.03.009

[30] Flores-Mireles, A., Hreha, T.N. and Hunstad, D.A., 2019. Pathophysiology, Treatment, and Prevention of Catheter-Associated Urinary Tract Infection. Topics in spinal cord injury rehabilitation, 25(3), pp.228-240.DOI: 10.1310/sci2503-228

[31] Pallett, A. and Hand, K., 2010. Complicated urinary tract infections: practical solutions for the treatment of multiresistant Gram-negative bacteria. Journal of antimicrobial chemotherapy, 65(suppl_3), pp.iii25-iii33. 10.1093/jac/dkq298 DOI: 10.1093/jac/dkq298

[32] Gomila, A., Carratalà, J., Eliakim-Raz, N., Shaw, E., Wiegand, I., Vallejo-Torres, L., Gorostiza, A., Vigo, J.M., Morris, S., Stoddart, M. and Grier, S., 2018. Risk factors and prognosis of complicated urinary tract infections caused by *Pseudomonas aeruginosa* in hospitalized patients: a retrospective multicenter cohort study. Infection and drug resistance, 11, p.2571.DOI: 10.2147/IDR.S185753

[33] Tellis, R.C., Moosabba, M.S. and Roche, R.A., 2017. Correlation between biofilm formation and antibiotic resistance in uropathogenic *Pseudomonas aeruginosa* causing catheter associated urinary tract infections. Eur. J. Pharm. Med. Res, 4, pp.248-252.

[34] Abbas, H.A., El-Ganiny, A.M. and Kamel, H.A., 2018. Phenotypic and genotypic detection of antibiotic resistance of *Pseudomonas aeruginosa* isolated from urinary tract infections. African health sciences, 18(1), pp.11-21. DOI: 10.4314/ahs.v18i1.3

[35] Anju, k.K., deepthy, b.J., suresh, g. And harish, p.V., 2019. A hospital based surveillance study of urinary tract infections caused by *Pseudomonas aeruginosa* and acinetobacter species with special emphasis on drug resistance. Paripex-indian journal of research, 8(12)..Doi: 10.36106.

[36] Saini, H., Vadekeetil, A., Chhibber, S. and Harjai, K., 2017. Azithromycin-ciprofloxacin-impregnated urinary catheters avert bacterial colonization, biofilm formation, and inflammation in a murine model of foreign-body-associated urinary tract infections caused by *Pseudomonas aeruginosa*. Antimicrobial agents and chemotherapy, 61(3), pp.e01906-16DOI: 10.1128/AAC.01906-16

[37] Newman, J.W., Floyd, R.V. and Fothergill, J.L., 2017. The contribution of *Pseudomonas aeruginosa* virulence factors and host factors in the establishment of urinary tract infections. FEMS Microbiology Letters, 364(15).DOI: 10.1093/femsle/fnx124

[38] Head, K., Chong, L.Y., Bhutta, M.F., Morris, P.S., Vijayasekaran, S., Burton, M.J., Schilder, A.G. and Brennan-Jones, C.G., 2020. Topical antiseptics for chronic suppurative otitis media. Cochrane Database of Systematic Reviews, (1).DOI: 10.1002/14651858. CD013055.pub2

[39] Mittal, R., Yan, D. and Liu, X.Z., 2016. *Pseudomonas aeruginosa* activates PKC-alpha to invade middle ear epithelial cells. Frontiers in microbiology, 7, p.255.DOI: 10.3389/fmicb.2016.00255

[40] Orji, F.T. and Dike, B.O., 2015. Observations on the current bacteriological profile of chronic suppurative otitis media in South eastern Nigeria. Annals of medical and health sciences research, 5(2), pp.124-128.DOI: 10.4103/2141-9248.153622

[41] Ali, S.M., Mahesar, J.H., Shahzad, J., Zaman, A., Sajid, T. and Khattak, S.K., 2020. Spectrum of *Pseudomonas aeruginosa* Sensitivity in Chronic Otitis Media. Journal of Saidu Medical College, 9(2).DOI:

[42] Jensen, R.G., Johansen, H.K., Bjarnsholt, T., Eickhardt-Sørensen, S.R. and Homøe, P., 2017. Recurrent otorrhea in chronic suppurative otitis media: is biofilm the missing link?. European Archives of Oto-rhino-laryngology, 274(7), pp.2741-2747. DOI: 10.1007/s00405-017-4586-8

[43] Sahu, M.C., Swain, S.K. and Kar, S.K., 2019. Genetically diversity of *Pseudomonas aeruginosa* isolated from chronic suppurative otitis media with respect to their antibiotic sensitivity pattern. Indian Journal of Otolaryngology and Head & Neck Surgery, 71(2), pp.1300-1308.DOI: 10.1007/s12070-018-1358-8

[44] Juyal, D., Sharma, M., Negi, V., Prakash, R. and Sharma, N., 2017. *Pseudomonas aeruginosa* and its sensitivity spectrum in chronic suppurative otitis media: A study from Garhwal hills of Uttarakhand State, India. Indian Journal of Otology, 23(3), p.180.DOI: 10.4103/indianjotol.INDIANJOTOL_31_14

[45] Ojo, A., Ojo, O., Adebajo, S., Oladotun, B., Sodunke, G. and Ejilude, O., 2020. In-Vitro Antibacterial Therapy of Ficus exasperata, Securinega virosa and *Tamarindus indica* Leaf Extract on Bacterial Isolate from Otitis Media effusion. Access Microbiology, 2(2), p.75.DOI:10.1099/acmi.fis2019.po0096.

[46] Knapen, D.G., Mulder, B., DeSouza, F., Linssen, G.C. and Veneman, T.F., 2017. *Pseudomonas aeruginosa* meningitis after visiting a swimming pool: a complicated dive. NETHERLANDS JOURNAL OF CRITICAL CARE, 25(1), pp.10-11.

[47] Saradna, A., Shankar, S., Ponnusamy, V., Shamian, B., Pendkar, C., Pascal, W. and Kupfer, Y., 2018. 733: *Pseudomonas aeruginosa* Meningitis With Massive Hemorrhagic Stroke. Critical CareMedicine, 46(1),p.353.DOI::10.1097/01.ccm.0000528747.40153.dd

[48] Salick, M.M., Faris, A.F., Wazir, A., D'souza, M. and Beegle, S., 2019. 622:

Index case of community-aquired MDR *Pseudomonas aeruginosa* meningitis. Critical Care Medicine, 47(1),p.291. DOI:10.1097/01.ccm.0000551374.77511.e8

[49] Pai, S., Bedford, L., Ruramayi, R., Aliyu, S.H., Sule, J., Maslin, D. and Enoch, D.A., 2015. *Pseudomonas aeruginosa* meningitis/ventriculitis in a UK tertiary referral hospital. QJM: An International Journal of Medicine, 109(2), pp.85-89.DOI: 10.1093/qjmed/hcv094

[50] Gallaher, C., Norman, J., Singh, A. and Sanderson, F., 2017. Community-acquired *Pseudomonas aeruginosa* meningitis. Case Reports, 2017, pp.bcr-2017.DOI: 10.1136/bcr-2017-221839

[51] Rodríguez-Lucas, C., Fernández, J., Martínez-Sela, M., Álvarez-Vega, M., Moran, N., Garcia, A., Menendez, C., García-Prieto, E. and Rodríguez-Guardado, A., 2020. *Pseudomonas aeruginosa* nosocomial meningitis in neurosurgical patients with intraventricular catheters: therapeutic approach and review of the literature. Enfermedades Infecciosas y Microbiología Clínica, 38(2), pp.54-58. DOI: 10.1016/j.eimc.2019.04.003

[52] Ye, J., Tan, L.H., Shen, Z.P., Yu, Y.S., Lai, D.M., Fan, J. and Shu, Q., 2020. Polymyxin for the treatment of intracranial infections of extensively drug-resistant bacteria in children after neurosurgical operation. World Journal of Pediatrics, pp.1-5.DOI: 10.1007/s12519-020-00350-8

[53] Jiang, L., Guo, L., Li, R. and Wang, S., 2017. Targeted surveillance and infection-related risk factors of nosocomial infection in patients after neurosurgical operation. Pakistan journal of pharmaceutical sciences, 30.

[54] Schaible, B., Crifo, B., Schaffer, K. and Taylor, C.T., 2020. The putative bacterial oxygen sensor *Pseudomonas* prolyl hydroxylase (PPHD) suppresses antibiotic resistance and pathogenicity in *Pseudomonas aeruginosa*. Journal of Biological Chemistry, 295(5), pp.1195-1201..DOI: 10.1074/jbc.RA119.010033.

[55] Ryder C, Byrd M, Wozniak DJ. Role of polysaccharides in *Pseudomonas aeruginosa* biofilm development. Current opinion in microbiology. 2007 Dec 1;10(6):644-8.

[56] Rodríguez-Lucas, C., Fernández, J., Martínez-Sela, M., Álvarez-Vega, M., Moran, N., Garcia, A., Menendez, C., García-Prieto, E. and Rodríguez-Guardado, A., 2020. *Pseudomonas aeruginosa* nosocomial meningitis in neurosurgical patients with intraventricular catheters: therapeutic approach and review of the literature. Enfermedades Infecciosas y Microbiología Clínica, 38(2), pp.54-58. DOI: 10.1016/j.eimc.2019.04.003

[57] Azeredo, J., Azevedo, N.F., Briandet, R., Cerca, N., Coenye, T., Costa, A.R., Desvaux, M., Di Bonaventura, G., Hébraud, M., Jaglic, Z. and Kačániová, M., 2017. Critical review on biofilm methods. Critical reviews in microbiology, 43(3), pp.313-351. .DOI:10.1080/1040841X.2016.1208146

[58] Cassin, E.K. and Tseng, B.S., 2019. Pushing beyond the envelope: the potential roles of OprF in *Pseudomonas aeruginosa* biofilm formation and pathogenicity. DOI: 10.1128/JB.00050-19.

[59] Maura, D. and Rahme, L.G., 2017. Pharmacological inhibition of the *Pseudomonas aeruginosa* MvfR quorum-sensing system interferes with biofilm formation and potentiates antibiotic-mediated biofilm disruption. Antimicrobial agents and chemotherapy, 61(12), pp.e01362-17. DOI: 10.1128/AAC.01362-17

[60] Pestrak, M.J., Baker, P., Dellos-Nolan, S., Hill, P.J., da Silva, D.P.,

Silver, H., Lacdao, I., Raju, D., Parsek, M.R., Wozniak, D.J. and Howell, P.L., 2019. Treatment with the *Pseudomonas aeruginosa* glycoside hydrolase PslG combats wound infection by improving antibiotic efficacy and host innate immune activity. Antimicrobial agents and chemotherapy, 63(6), pp.e00234-19.DOI: 10.1128/AAC.00234-19.

[61] Wei, Q. and Ma, L.Z., 2013. Biofilm matrix and its regulation in *Pseudomonas aeruginosa*. International journal of molecular sciences, 14(10), pp.20983-21005.DOI: 10.3390/ijms141020983

[62] Ghazalibina, M., Morshedi, K., Farahani, R.K., Babadi, M. and Khaledi, A., 2019. Study of virulence genes and related with biofilm formation in *Pseudomonas aeruginosa* isolated from clinical samples of Iranian patients; A systematic review. Gene Reports, p.100471.DOI: 10.1016/j.genrep.2019.100471.

[63] Winstanley C, O'Brien S, Brockhurst MA. *Pseudomonas aeruginosa* evolutionary adaptation and diversification in cystic fibrosis chronic lung infections. Trends in microbiology. 2016 May 1;24(5):327-37.

[64] Chatterjee, M., Anju, C.P., Biswas, L., Kumar, V.A., Mohan, C.G. and Biswas, R., 2016. Antibiotic resistance in *Pseudomonas aeruginosa* and alternative therapeutic options. International Journal of Medical Microbiology, 306(1), pp.48-58.DOI: doi.org/10.1016/j.ijmm.2015.11.004

[65] Thirumalmuthu, K., Devarajan, B., Prajna, L. and Mohankumar, V., 2019. Mechanisms of Fluoroquinolone and Aminoglycoside Resistance in Keratitis-Associated *Pseudomonas aeruginosa*. Microbial Drug Resistance, 25(6), pp.813-823.DOI: 10.1089/mdr.2018.0218

[66] Falagas, M.E., Koletsi, P.K. and Bliziotis, I.A., 2006. The diversity of definitions of multidrug-resistant (MDR) and pandrug-resistant (PDR) Acinetobacter baumannii and *Pseudomonas aeruginosa*. Journal of medical microbiology, 55(12), pp.1619-1629.DOI: 10.1099/jmm.0.46747-0.

[67] Siempos, I.I., Vardakas, K.Z., Manta, K.G. and Falagas, M.E., 2007. Carbapenems for the treatment of immunocompetent adult patients with nosocomial pneumonia. European Respiratory Journal, 29(3), pp.548-560. DOI: 10.1183/09031936.00080206

[68] Tsao, L.H., Hsin, C.Y., Liu, H.Y., Chuang, H.C., Chen, L.Y. and Lee, Y.J., 2018. Risk factors for healthcare-associated infection caused by carbapenem-resistant *Pseudomonas aeruginosa*. Journal of microbiology, immunology and infection, 51(3), pp.359-366. DOI: 10.1016/j.jmii.2017.08.015

[69] Balasubramanian, D., Schneper, L., Kumari, H. and Mathee, K., 2013. A dynamic and intricate regulatory network determines *Pseudomonas aeruginosa* virulence. Nucleic acids research, 41(1), pp.1-20.DOI: 10.1093/nar/gks1039

Chapter 3

Pseudomonas aeruginosa as a Cause of Nosocomial Infections

Silvia Labovská

Abstract

Pseudomonas aeruginosa, as a gram-negative aerobic rod, is still one of the most resistant agents of nosocomial infections. It is used for the development of respiratory, urinary and wound infections. It causes bacteremia, especially in patients who are hospitalized for anesthesiology and resuscitation department or ICU, who often have respiratory insufficiency and hemodynamic instability and require artificial lung ventilation. Mechanical ventilation itself is a significant risk factor for the development of pseudomonad pneumonia. *Pseudomonas aeruginosa* has enzymes that are encoded on both chromosomes and plasmids, often in combination with other mechanisms of resistance, such as reducing the permeability of the outer or cytoplasmic membrane. Due to carbapenemases, *Pseudomonas aeruginosa* loses sensitivity to carbapenem and becomes resistant to this antibiotic. It also becomes resistant to aminoglycosides, cephalosporins and ureidopenicillins. It is also resistant to Quaternary disinfectants. The reservoir of pseudomonas nosocomial infection is hospital water, taps, shower roses, swimming pools, healing waters and others. The intervention of anti-epidemic measures in the case of infections caused by pseudomonad strains has not yet reached such sophistication as in the case of MRSA for time, personnel and economic reasons. In the absence of an epidemic, intervention in sporadic cases consists of informing nursing staff of the occurrence of a multidrug-resistant agent, including providing all patient demographics and relieving careful adherence to the barrier treatment, cleansing, disinfection and isolation regimen.

Keywords: nosocomial infections, respiratory insufficiency, pnemonia, lung ventilation, resistance

1. Introduction

Nosocomial infections (NI) are a global problem in hospital care. This is a significant complication that worsens the prognosis of the underlying disease, increases mortality, prolongs hospitalization, worsens the quality of life of patients and increases the cost of treatment, so NI pays special attention. In the United States, the Center for Disease Control and Prevention (CDC) is the leading authority in this area. CDC procedures and guidelines are the most widely used standard worldwide. In the USA, a nationwide NI surveillance system has been organized since the 1970s. The United Kingdom also has a long tradition in the control of NI, which is organized in the system of laboratory service of the British public health service (Public Health Laboratory Service). The main guarantor, which organizes congresses dedicated to NI, is the Hospital Infection Society (HIS), which publishes a globally important and recognized journal - the Journal of Hospital Infection [1]. The aim of the journal is to publish high-quality research and information related to the prevention and control of NI [2]. NIs need to be diagnosed and treated in time, but the most important thing is their prevention in various hospital wards, especially in intensive care units. Infections acquired in connection with hospitalization can lead to significant morbidity and mortality, but preventive anti-infective measures can significantly affect these results. Equally important is prevention in hospital staff in order to reduce the risk of infections spreading to other patients and staff. In this way, it is possible to prevent the absence of staff from work, which can have a positive effect on the skills of the staff of the intensive care unit. Nosocomial infections are also associated with financial expenses, which include hospital expenses, reduced productivity of sick staff as well as their income due to absence from work.

Pseudomonas aeruginosa, as a gram-negative aerobic rod, is still one of the most resistant agents of nosocomial infections. *P. aeruginosa* causes 10-11% of all NI. This result is due to the resistance of this microorganism to desinfectants and many antimicrobials. It is involved in the development of respiratory, urinary and wound infections. It causes bacteremia, especially in patients who are hospitalized at anesthesiology and resuscitation department or ICU, who often have respiratory insufficiency and hemodynamic instability and require artificial lung ventilation. Mechanical ventilation itself is a significant risk factor for the development of pseudomonad pneumonia. *P. aeruginosa* has enzymes that are encoded on both chromosomes and plasmids, often in combination with other mechanisms of resistance, such as reducing the permeability of the outer or cytoplasmic membrane. Due to carbapenemases, *P. aeruginosa* loses sensitivity to carbapenem and becomes resistant to this antibiotic. It also becomes resistant to aminoglycosides, cephalosporins and ureidopenicillins. It is also resistant to Quaternary desinfectants. The reservoir of pseudomonas nosocomial infection is hospital water, taps, shower roses, swimming pools, healing waters and others. It occurs in sinks, humidifiers, anesthesia machines, inhalers, hand brushes and other places that meet suitable conditions, which means ambient humidity. Pseudomonads contaminate lubricating gels and disinfectants [3].

The intervention of anti-epidemic measures in the case of infections caused by pseudomonad strains has not yet reached such sophistication as in the case of MRSA for time, personnel and economic reasons. In the absence of an epidemic, intervention in sporadic cases consists of informing nursing staff of the occurrence of a multidrug-resistant agent, including providing all patient demographics and relieving careful adherence to the barrier treatment, cleansing, desinfection and isolation regimen [3].

If we add eye, ear, nose, and throat infections to pneumonia, then respiratory tract infections are the most common site of nosocomial infections for almost all age groups in pediatric JIS [4]. Much attention has been paid to ventilator-associated pneumonia (VAP) as the most common and potentially preventable nosocomial infection. Other nosocomial respiratory infections include sinusitis, otitis media and tracheitis. Contamination of the patient's respiratory tract may come from a device with which the patient has been in direct contact, namely an endotracheal tube, nasogastric tube, aspiration catheters, bronchoscopes, but also from a device with which he has not been in direct contact, such as a mechanical ventilator, ventilator hose, nebulizers and devices that supply oxygen. The human vector that most likely transmits infection to a patient is hospital staff. The most common risk factors are poor hand hygiene, insufficient isolation of patients and contaminated objects such as stethoscopes. Family members and other patients may also transmit the infection to patients hospitalized in a pediatric ICU. All of these factors must be considered and controlled to minimize the occurrence of nosocomial respiratory tract infections [2].

2. Ventilator-associated pneumonia

Nosocomial pneumonia is the second most common nosocomial infection in pediatric ICUs after catheter infections of the bloodstream. Nosocomial infection can occur in any patient, but is most common in infants, young children, and patients over 65 years of age. Patients in pediatric ICUs who are most at risk for pneumonia are patients who have been intubated and mechanically ventilated. The risk increases due to the circumvention and alteration of the host's defenses, as the vocal cords remain open and the risk of aspiration of gastrointestinal contents increases. The risk of nosocomial pneumonia is 6-20 times higher in ventilated patients compared to non-ventilated patients. Ventilator-associated pneumonia (VAP) is defined as the development of new pneumonia for at least 48 hours after the start of mechanical ventilation. Independent risk factors for the development of VAP in children are immunodeficiency, immunosuppression and neuromuscular blockade. Other risk factors are genetic syndromes with neuromuscular weakness, burns, steroid administration and total parenteral nutrition [2]. Children have a higher risk of VAP with antibiotics, with a longer stay in the ICU, with catheters in place with a risk of haematogenous spread, treatment with H2-receptor blockers, reintubation and transport outside the ICU during intubation. The presence of a nasogastric tube increases the risk as it provides a direct pathway from the upper gastrointestinal tract to the oropharynx. In-line nebulizers and manipulation of the ventilator circuit can affect the risk of nosocomial pneumonia. VAP in children accounts for 10-26% of nosocomial infections. The incidence of pediatric nosocomial pneumonia within the hospital is highest at the neonatal JIS, followed by the pediatric JIS, and the pediatric ward. Nosocomial pneumonia has the highest mortality of all pediatric nosocomial infections and ranges from 20–70%. Although the duration of endotracheal intubation increases the risk of nosocomial pneumonia, the highest risk is during the first 2 weeks of intubation. Almost all intubated children have a colonized endotracheal tube with nosocomial microorganisms within 5 days [2]. The most frequently identified bacteria in pediatric JIS are gram-negative bacilli, especially *P. aeruginosa*. Mortality is higher with gram-negative microorganisms. *P. aeruginosa* is one of the leading causes of ventilator-associated pneumonia (VAP) in the US and Europe [5–7]. VAP due to *P. aeruginosa* is increasing in incidence and poses unique challenges for its clinical management.

2.1 Symptoms and diagnosis of VAP

The diagnosis of VAP in children can be made on a clinical basis without the use of bronchoscopy. A set of clinical diagnostic criteria and alternative criteria that vary with age are given in the table (**Table 1**). The presence of pneumatoceles on chest X-rays in children under 12 months of age meets the radiographic criteria for pneumonia, which are listed in the table. The diagnosis of VAP can be made based on clinical and radiographic criteria. Identification of the causative microorganism is essential for targeted antibiotic therapy. Identification of the microorganism is difficult because endotracheal tube culture is inaccurate due to colonization of the endotracheal tube and upper airways by gram-negative bacilli and staphylococci, which occurs within a few days after intubation. In adult and older children, bronchoalveolar lavage and protected swab specimens have been used successfully. In young children, it is not possible to obtain a protective sample for the size of the required bronchoscope, and the bronchoalveolar lavage performed has a high incidence of contamination. Methods for determining the causative microorganism are positive blood culture that cannot be explained by other sources, positive pleural fluid cultures, and a positive bronchoalveolar lavage sample despite its limitations, >5% of bronchoalveolar lavage cells containing intracellular bacteria and positive

	All patients	1-12 year of age	<12 months of age
Chest film	At least 2 serial CXR with new or progressive and persistent infiltrate or consolidate or cavitation that developes later than 48 hrs post initiation of mechanical ventilation		
Additional Criteria	At least one of shaded criteria AND At least two of the non-shaded criteria	At least 3 of the criteria below	Worsening gas exchange AND at least 3 of the criteria below
Temperature	>38°C without other recognized cause	>38,4°C or <37°C without other recognized cause	Temperature instability without other recognized cause
WBC count	<4000/mm3 OR ˃12,000/ mm3	<4000/mm3 OR ˃>15,000/mm3	˂4000/mm3 OR ˃15,000/mm3 and band forms ˃10%
Altered mental status	If >70 years of age without other recognized cause	Not applicable	Not applicable
Sputum/ Secretions	New onset purulent sputum OR change in character of sputum OR increased respiratory symptoms		
Respiratory Symptoms	New onset or worsening of cough, dyspnea, or tachypnea		Apnea, tachypnea, increased work of breathing, or grunting
Auscultation findings	Rales or bronchial breath sounds		Wheezing, rales, or ronchi
Cough	Not applicable as separate criteria		+
Worsening oxygenation or ventilation	Present	Present	Required criteria
Heart rate	Not applicable		<100 beats/min OR > 170 beats/min

Table 1.
Clinical criteria for diagnosing VAP by age [2].

pulmonary parenchyma culture. When nosocomial pneumonia is suspected, empirical treatment should be initiated to cover the most likely microorganisms, taking into account hospital resistance. Once the agent is identified, the antibiotic coverage needs to be adjusted [2].

2.2 Prevention of VAP

In 2004, The Institute for Healthcare Improvement developed a set of evidence-based recommendations for practitioners to reduce mortality. The evidence was based on research in adults.

The package of recommendations for VAP in adults includes the following interventions [8]:

a. raising the patient's head above the bed between 30 and 45°,

b. a break in sedation and daily reassessment of extubation,

c. prophylaxis of stress ulcers,

d. prophylaxis of deep vein thrombosis.

The application of these measures can reduce the incidence of VAP to 45%, although the last 2 points do not directly lead to nosocomial pneumonia, but are designed to treat complications in monitored, sedentary adult patients with ICU. In children, many centers use only low-risk interventions such as raising the head above the bed, considering extubation, and using stress ulcer prophylaxis. Intervention such as omission of sedation is unpredictable and risky in young children due to the high risk of unwanted extubation [9].

Measures often used in pediatric centers focus on specific risk factors [2]:

- measures to prevent iatrogenic spread of infection compliance with good hand hygiene use of general preventive measures use of appropriate isolation techniques according to infectious microorganisms

- measures to prevent aspiration of gastric contents elevated head above bed between 30 and 45 degrees monitor/drainage of gastric contents

- measures to improve oral hygiene mouthwashes/cleaning with chlorhexidine 0.12% use of toothbrush and oral swab in daily oral hygiene

- measures to reduce risk factors of the endotracheal tube use of in-line suction device, where is suitable and available preferential suction of the hypopharynx over endotracheal suction and relocation of the ET tube

- measures to prevent contamination of respiratory equipment single-purpose oropharyngeal suction device prevention of condensate accumulation in the respiratory circuit prevention of contamination of respiratory device

- measures to reduce the length of mechanical ventilation daily consideration of extubation attempts interruption of neuromuscular blockade.

Hygiene of hands with alcoholic solutions or soap and water, together with adherence to general precautions and appropriate isolation, are the most effective methods. The raised position of the head prevents aspiration of the stomach contents. The risk of aspiration can be further minimized by decompression of the stomach with a gastric tube and continuous monitoring of the residue. Mouth hygiene is important. The American Dental Association recommends starting continuous oral hygiene in infants before the appearance of dentition. The recommendation for the use of oral swabs and brushing teeth in critically ill patients is based on the fact that the dental plaque consists predominantly of gram-negative bacteria and forms within 48 hours of admission to the ICU [2].

In children, secretion of secretions from the hypopharynx is recommended to prevent VAP. It is recommended that this aspiration be performed prior to aspiration from the endotracheal tube, to prevent aspiration of secretions from the hypopharynx, and prior to manipulation of the endotracheal tube. In some centers, they also aspirate secretions before positioning the patient on the bed. The use of a closed in-line extraction system may not have a direct effect on reducing the incidence of VAP, but may be effective in preventing contamination of the extraction device. Condensed steam in the respiratory circuit can potentially contaminate and theoretically cause infection, so condensate must be removed from the circuit. Staff should be conscientious and avoid contaminating the respirator and its accessories [9].

In the prevention of nosocomial pneumonia, it is important to minimize the length of the patient's mechanical ventilation. The presence of an endotracheal tube poses a risk of VAP and not the positive pressure ventilation associated with it.

Daily consideration is recommended as to whether the patient can be extubated. Discontinuation of sedation is impractical for most children in pediatric ICUs, as it can potentially lead to unwanted extubation, especially in children who are small enough to cooperate or understand the need for intensive care interventions. Studies in adults and children show that the use of non-invasive ventilation in ICU contributes to reducing the incidence of VAP [2].

3. Treatment

One of the most important challenges for physicians is the adequate treatment of infections due to Gram-negative pathogens because of the increasing antimicrobial resistance in the healthcare setting [10].

Among infections caused by Gram-negative rods, *P. aeruginosa* has a leading role [11], especially in critically ill and immunocompromised patients. Antimicrobial resistance has led to a serious restriction in treatment options for *P. aeruginosa* infections.

An anti-pseudomonal cephalosporin, or a carbapenem, or an anti-pseudomonal β-lactam/BLI represents potential options for definitive therapy. Aminoglycosides should not be used as monotherapy because success rates for aminoglycosides are low [8]. This may be due to the poor penetration of aminoglycosides into the lung, which require high peak serum concentrations to obtain adequate lung concentrations, thus increasing the risk of nephrotoxicity or ototoxicity [12, 13]. However, because in Europe fluoroquinolone resistance rate in P. aeuruginosa exceeds 30% [14], it is appropriate to use combination therapy including aminoglycosides for empirical therapy of serious VAP. A based approach is recommended of the prescription of an anti-pseudomonal β-lactam (piperacillin/tazobactam, ceftolozane/tazobactam, ceftazidime, cefepime, or a carbapenem) plus a second anti-pseudomonal agent (aminoglycoside or a fluoroquinolones). As for aerosol therapy, there is not routinely recommended the use of inhaled antibiotics for the treatment of *P. aeruginosa* VAP. However, they may be considered as an adjunctive to intravenous therapy in cases of infections due to MDR (Multi-drug resistance) strains [15].

4. Factors of nosocomial infections

Factors that affect the complex process of origin and spread of nosocomial infections are divided into internal and external:

- internal factors are closely related to the biological balance of the patient: age (over 60 years, up to 3 years), alcoholism, drug addiction, hormonal disorders (diabetes), malignant tumors, immunodeficiency, obesity, malnutrition, circulatory disorders, polytraumas, burns, pressure ulcers, ulcus cruris, other serious diseases (liver disease, AV shunt, cardiomyopathy),

- external factors are related to therapeutic, prophylactic and diagnostic interventions and are used exclusively in treatment of patients in hospital facilities: length of hospital stay, surgery, transplantation, tracheostomy, endotracheal cannula, gastric tube, urinary catheterization, iv catheterization, infusion, transfusion, foreign bodies, drainage, instrumental procedure, repeated anesthesia, endoscopy, hemodialysis, radiation therapy, cytostatic therapy, immunosuppressive therapy, broad spectrum ATB therapy, hormonal therapy [3].

Hospital placement: plays an important role, with the highest incidence being typical of ICU. The incidence of nosocomial infection also depends on the type of ICU, while the different incidence will be on surgical, traumatological, burn, neurological, neurosurgical or cardiological ICU. Pediatric ICU is unique in that it provides care in all of these areas for all children except newborns [2].

The patient's age may affect the risk of nosocomial infection. In the pediatric population, young children are most at risk, especially newborns. The highest incidence of nosocomial infections among pediatric patients is in children less than 1 year of age. The relative immaturity of the newborn's immune system, associated with routine ICU procedures that bypass the physical barriers of infection such as skin and mucous membranes, is responsible for the increased risk. Parenteral nutrition with high concentrations of glucose and lipids is another risk factor for infection. The fact that premature infants are most affected by these risk factors explains why neonatal ICUs have a higher incidence of nosocomial infections than pediatric ICUs [2, 4, 16]. Pediatric ICU is also unique in that each childhood has a different incidence depending on the type of nosocomial infection. In children under 5 years of age, the 3 most common nosocomial infections are in the following order: bloodstream infections, so-called bloodstream infections, pneumonia and urinary tract infections. In children aged 5 to 12 years, the 3 most common are nosocomial infections: pneumonia, bloodstream infections and urinary tract infections. In adolescents, the order of the most common nosocomial infections is: bloodstream infections, urinary tract infections, and then pneumonia [2, 4, 16]. Immunosuppressed patients after chemotherapy, human immunodeficiency virus infection, or steroid use are equally at risk for developing nosocomial infection.

Nosocomial infections do not have an apparent sex predilection.

Particular risk factors for the development of nosocomial infection are length of hospital stay and initial antibiotic therapy [17].

Staff shortages are a particular risk factor for the increased incidence of nosocomial infections, due to increased staff workload and poor hand hygiene [18].

Erythrocyte transfusion is a risk factor for the development of nosocomial infections in critically ill patients on ICU. In a prospective study, the incidence of nosocomial infection was 14.3% in patients with blood transfusions and 5.8% in patients without blood transfusions. In the group of patients with blood transfusions, there was a higher incidence of nosocomial infections, which was significant in seriously ill patients with a probability of survival of less than 25%. Patients with more than a 25% chance of survival had higher mortality, longer stays on the ICU, and longer hospitalizations compared with patients who did not receive a blood transfusion [19].

5. Prevention

It is not possible to eliminate all nosocomial infections, but one third of cases could be prevented if organized infection control programs were put in place.

Preventive measures can be divided into 2 categories, namely standard measures and transmission-based measures. Standard measures can always be used and are designed to prevent personnel from coming into contact with potentially infectious body fluids. The most important standard measure is hand hygiene. Washing hands with soap and water is considered the gold standard. The use of anhydrous antiseptic agents is accepted, but not in cases where visible dirt is present, proteinaceous body fluids such as blood, or spores contamination is suspected. In these cases, it is necessary to use soap and water. Hand hygiene must be observed before and after the patient's examination, but also when gloves are worn. In case of contact with body fluids or secretions, it is advisable to use barriers such as gloves, masks, eye protectors and coats [2, 3].

Transmission-based measures aim to protect against the transmission of infectious micro-organisms from patients with a proven or suspected infection, as well as from patients colonized by specific micro-organisms. These additional measures are more than standard measures and are based on the path of transmission: contact, droplets, or airborne transmission.

Contact transfer measures apply to a wide range of micro-organisms that spread by direct contact with the patient or by indirect contact through contaminated objects such as toys, a stethoscope and unwashed hands. Preventive measures include, in addition to standard measures, isolation rooms for the patient or group, coats and gloves.

Droplet transfer measures are directed against microorganisms that spread a short distance from the patient by coughing and sneezing. These measures include isolation rooms for one patient or for a group of patients with the same microorganism. Healthcare professionals should wear masks with eye protection in addition to standard measures.

Measures to prevent airborne transmission include additional precautions against microorganisms which spread through the air stream. Patients should be isolated in rooms with ionized air. For other airborne microorganisms, a respirator is required when entering the patient's room. Isolation of a patient may be based on clinical symptoms or circumstances present on admission to the hospital and should always be initiated before isolating the microorganism [2, 3].

6. Conclusions and further directions

Nosocomial infections are a major cause of morbidity and mortality in the intensive care unit, which can usually be prevented. Although not all nosocomial infections can be eliminated due to the specific nature of JIS patients, the incidence can be significantly reduced by control measures. Simple measures such as strict hand hygiene, isolation, sterility, elevated head position, judicious use, and prompt removal of central catheters, urinary catheters, and endotracheal tubes can dramatically affect the frequency of nosocomial infections. Although not all nosocomial infections can be prevented, intensive care targets should be zero. Consistent application and monitoring of the effectiveness of infection control measures must go a long way towards achieving the goal. The medical community must take steps to reduce and prevent nosocomial infections. Future efforts should be made to distinguish community-acquired infection from nosocomial infection, to reduce the development of resistant organisms through prudent use of antibiotics, to design JIS to isolate patients and ensure hand hygiene, and to develop barrier design [2, 3].

6.1 Further directions

- Adherence to infection control procedures, including hand hygiene, is one of the most useful and well-established methods for preventing nosocomial infections.

- Isolation measures are crucial in preventing the transmission of infections among hospitalized patients.

- Following cultures at the time of patient admission may reduce the spread of nosocomial resistant organisms.

DOI: http://dx.doi.org/10.5772/intechopen.95908

- Prevention of catheter blood infections begins at the time of insertion with sterility. Catheter care and the use of catheters that are impregnated with antiseptics or antibiotics may further reduce the risk of infection.

- Routine removal of central venous catheters does not reduce the risk of catheter blood infections.

- VAP prevention is facilitated by the use of a protocol that includes raising the head above bed level and considering extubation daily.

Author details

Silvia Labovská
Clinic for Children and Adolescents, Teaching Hospital Nitra, Slovakia

*Address all correspondence to: silvialabovska@gmail.com

References

[1] Gray J: Journal of Hospital Infection, 2016. In www.journal.elsevier.com/journal-of-hospital-infection ISSN: 0195-6701.

[2] Straumanis JP: Nosocomial infections in the pediatric intensive care unit. In: *Pediatric Intensive Care.* 2008: 1400-1418, ISBN-13: 978-0-7817-8275-3.

[3] Šramova H, et al: Nozokomiální nákazy. Praha, Maxdorf, 2013: 400 s. ISBN 978-80-7345-286-5.

[4] Richards MJ, Edwards JR, Culver DH, et al.: Nosocomial infections in pediatric intensive care units in United States. National Nosocomial Infections Surveillance System. In: *Pediatrics.* 1999, 4: 103. http://pediatrics.aappublications.org/ content/ 103/ 4/ e39.full.pdf+html ISSN: 1098 - 4275.

[5] Fernandez-Barat L, Ferrer M, De Rosa F, Gabarrus A, Esperatti M, Terraneo S, Rinaudo M, Li Bassi G, Torres A. Intensive care unit-acquired pneumonia due to *Pseudomonas aeruginosa* with and without multidrug resistance. J Infect. 2017;74[2]:142-52. http://dx.doi.org/10.1016/j.jinf.2016.11.008. [PubMed] [Google Scholar]

[6] Koulenti D, Tsigou E, Rello J. Nosocomial pneumonia in 27 ICUs in Europe: perspectives from the EU-VAP/CAP study. Eur J Clin Microbiol Infect Dis. 2017;36[11]:1999-2006. http://dx.doi.org/10.1007/s10096-016-2703-z. [PubMed] [Google Scholar]

[7] Weber DJ, Rutala WA, Sickbert-Bennett EE, Samsa GP, Brown V, Niederman MS. Microbiology of ventilator-associated pneumonia compared with that of hospital-acquired pneumonia. Infect Control Hosp Epidemiol. 2007;28[7]:825-31. http://dx.doi.org/10.1086/518460. [PubMed] [Google Scholar]

[8] Kalil AC, Metersky ML, Klompas M, Muscedere J, Sweeney DA, Palmer LB, Napolitano LM, O'Grady NP, Bartlett JG, Carratala J, El Solh AA, Ewig S, Fey PD, File TM, Jr, Restrepo MI, Roberts JA, Waterer GW, Cruse P, Knight SL, Brozek JL. Management of adults with hospital-acquired and ventilator-associated pneumonia: 2016 clinical practice guidelines by the Infectious Diseases Society of America and the American Thoracic Society. Clin Infect Dis. 2016;63[5]:e61–e111. http://dx.doi.org/10.1093/cid/ciw353. [PMC free article] [PubMed] [Google Scholar]

[9] Curley MA, Schwalenstocker E, Deshpande JK, et al.: Tailoring the Istitute for Health Care Improvement 100,100 Lives Campaign to pediatric settings: The example of ventilator-associate pneumonia. In: *Pediatr Clin North Am.* 2006; 53 [6] 1231-1251.

[10] Pena C, Suarez C, Tubau F, Dominguez A, Sora M, Pujol M, Gudiol F, Ariza J. Carbapenem-resistant *Pseudomonas aeruginosa*: factors influencing multidrug-resistant acquisition in non-critically ill patients. Eur J Clin Microbiol Infect Dis. 2009;28[5]:519-22. http://dx.doi.org/10.1007/s10096-008-0645-9. [PubMed] [Google Scholar]

[11] El Zowalaty ME, Al Thani AA, Webster TJ, El Zowalaty AE, Schweizer HP, Nasrallah GK, Marei HE, Ashour HM. *Pseudomonas aeruginosa*: arsenal of resistance mechanisms, decades of changing resistance profiles, and future antimicrobial therapies. Future Microbiol. 2015;10[10]:1683-706. http://dx.doi.org/10.2217/fmb.15.48. [PubMed] [Google Scholar]

[12] Carcas AJ, Garcia-Satue JL, Zapater P, Frias-Iniesta J. Tobramycin penetration into epithelial lining fluid of patients with pneumonia. Clin Pharmacol

Ther. 1999;65[3]:245-50. http://dx.doi.org/10.1016/S0009-9236(99)70103-7. [PubMed] [Google Scholar]

[13] Levy J, Baran D, Klastersky J. Comparative study of the antibacterial activity of amikacin and tobramycin during Pseudomonas pulmonary infection in patients with cystic fibrosis. J Antimicrob Chemother. 1982;10[3]:227-34. http://dx.doi.org/10.1093/jac/10.3.227. [PubMed] [Google Scholar]

[14] Slekovec C, Robert J, Trystram D, Delarbre JM, Merens A, van der Mee-Marquet N, de Gialluly C, Costa Y, Caillon J, Hocquet D, Bertrand X ONERBA. *Pseudomonas aeruginosa* in French hospitals between 2001 and 2011: back to susceptibility. Eur J Clin Microbiol Infect Dis. 2014;33[10]:1713-7. http://dx.doi.org/10.1007/s10096-014-2125-8. [PubMed] [Google Scholar]

[15] Bassetti M, Vena A, Croxatto A, Righi E, Guery B. How to manage Pseudiomonas aeruginosa infections. Drugs Context. 2018; 7. https://www.ncbi.nlm.nih.gov/pmc/articles/PMC5978525/ doi: 10.7573/dic.212527.

[16] Rosenthal VD, Maki DG, Salomao R, et al.: Device-associated nosocomial infections in 55 intensive care units of 8 developing countries. In: *Ann Intern Med.* 145, 2006, 145[8]: 582-591.

[17] Arantes A, Carvalho E, Medeiros EA, et al: Pediatric risk of mortality and hospital infection. In: *Infect Control Hosp Epidemiol.* 2004; 25[9]: 783-785.

[18] Coffin SE, Zaoutis TE: Infection control, hospital epidemiology, and patient safety. In: *Infect Dis Clin North Am.* 2005; 19[3]: 647-665.

[19] Taylor RW, Obrien J, Tŕottier SJ, et al.: Red blood cell transfusions and nosocomial infections in critically ill patients. In: *Crit Care Med.* 2006, 34 [9]: 2302.

Chapter 4

Pseudomonas aeruginosa Biofilm Lung Infection in Cystic Fibrosis: The Challenge of Persisters

Gianmarco Mangiaterra, Mehdi Amiri, Nicholas Cedraro and Francesca Biavasco

Abstract

Pseudomonas aeruginosa lung infection is difficult to eradicate due to the multiple (intrinsic and acquired) antibiotic resistance of bacteria and to their ability to produce a thick biofilm. Antibiotic treatment is hampered by poor antibiotic diffusion, efflux pump overexpression and the development of a persistent subpopulation with low metabolic activity. This is a cause for special concern in Cystic Fibrosis (CF) patients, where *P. aeruginosa* lung infection is the chief cause of morbidity and mortality. Combined tobramycin-ciprofloxacin treatment is routinely adopted due to the low frequency of resistant strains and its ostensible ability to control the infection. Nevertheless, symptoms usually recur, mainly due to the antibiotic persisters, which are difficult to detect in routine cultural microbiological assays. This chapter describes the issues involved in the microbiological diagnosis of *P. aeruginosa* lung infection in CF patients and the possible role of subinhibitory antibiotic concentrations in persister development and infection recurrence.

Keywords: *Pseudomonas aeruginosa*, biofilms, antibiotic resistance, bacterial persisters, viable but non-culturable forms, infection recurrence

1. Introduction

Infectious biofilms have long been recognized as a severe clinical problem due to their tolerance to antimicrobials and their successful evasion of host defenses [1]. Their eradication is hampered by a variety of factors that are related to the sessile lifestyle and high cell density typical of biofilms, chiefly the poor diffusion of antibiotics and immune cells, the selection of antibiotic-resistant mutants, the development of intrinsic antibiotic-resistant phenotypes, like small colony variants (SCVs), and the spread of resistance genes among the bacterial populations through horizontal gene transfer (HGT) events. The problem is compounded by the development of antibiotic-unresponsive dormant cells, which upon reaching the late stage of dormancy can become non-culturable [2–4], thus escaping detection by routine culture-based assays [5, 6]. The difficulty of eradicating bacterial biofilms is a key factor in recurrent and chronic infections [1, 7, 8].

The opportunistic pathogen *Pseudomonas aeruginosa* is one of the bacteria most frequently involved in biofilm-related infections. Although most strains are environmental, the pathogen can live in symbiosis with a variety of hosts including plants, insects and animals. In humans it is an important nosocomial pathogen responsible for a variety of infections that have a strong tendency to recur, particularly in burn patients and in those with lung involvement. Like other opportunistic pathogens, it typically affects immunocompromised individuals [9]. However, the subjects most prone to develop *P. aeruginosa* infection are patients with cystic fibrosis (CF).

2. *P. aeruginosa* biofilms and lung infection in cystic fibrosis

2.1 *P. aeruginosa* biofilms and antibiotic resistance

The ubiquitous presence of *P. aeruginosa*, its prevalence and persistence in clinical settings and its intrinsic resistance to therapeutics are underpinned by an extraordinary arsenal of response mechanisms [10]. In particular, bacteria are protected by biofilms from adverse environmental conditions like phagocytosis, oxidative stress, nutrient/oxygen restriction, metabolic waste accumulation and antimicrobial agents [1, 11]. The matrix – which provides a favorable niche for intense cell–cell interaction and communication and a reservoir of metabolic substances, nutrients and energy [12] – accounts for 90% of the dry weight of the biofilm mass. Its main constituents are extracellular polysaccharides, proteins, extracellular DNA (eDNA), lipids, especially rhamnolipids, and other secreted molecules, such as the siderophores pyoverdine and pyocheclin, pyocyanin and phenazines. The production of all these components is highly regulated by *quorum sensing* (QS). *P. aeruginosa* biofilm development is characterized by the production of large amounts of three types of extracellular polysaccharides: Psl, Pel and alginate. Psl and Pel are the main constituents of the extracellular matrix in non-mucoid strains and are involved in the early stages of biofilm formation and in cell–cell interactions, whereas alginate overproduction is associated with the mucoid phenotype, the hallmark of chronic infection, and is indicative of disease progression and long-term persistence.

Biofilm development is held to be a differentiation process – activated in response to a variety of environmental stimuli – that alters pathogen behavior and results in the adoption of a sessile lifestyle [13]. Biofilms are characterized by an intricate regulation network that induces the development of different bacterial subpopulations and the emergence of antibiotic-resistant variants, which are a typical trait of *P. aeruginosa* biofilms [14]. The heterogeneity of the biofilm bacterial population is associated with the presence of niches with distinctive environmental characteristics that modulate gene expression patterns [15].

Biofilm formation is regulated by a number of redundant mechanisms of which QS is the most widely investigated. Four different QS systems, Las, Rhl, Pqs and Iqs, each characterized by a specific signal molecule and a receptor protein, have been described in *P. aeruginosa*. QS systems are involved in the regulation of several metabolic and pathogenic pathways that have a significant role in bacterial fitness in the environment as well as in the host. Their interplay is governed by a complicated hierarchical network, where the Las system directly regulates the Pqs and the Rhl systems [15].

Additional regulator systems, which sense the changes in the extracellular environment and regulate gene expression accordingly, also seem to be key factors in biofilm population dynamics. The best known is the Gac/Rsm system, which is the main factor controlling the switch from the planktonic to the sessile lifestyle in *P. aeruginosa* [13]. It encompasses two proteins, GacS/GacR, which sense and respond to environmental stimuli, promoting the synthesis of two small RNAs, RsmZ and RsmY, which bind and sequester the post-transcriptional regulator RsmA [16]. It induces the expression of virulence factors and of other genes playing roles in colonization and acute infection processes, such as the genes involved in motility (synthesis of pili) and in the type III secretion system; at the same time, it represses some genes implicated in chronic infections, such as those encoding the production of alginate and other exopolysaccharides, which constitute the biofilm matrix. RsmA sequestration seems to be a central mechanism in the shift from the planktonic to the biofilm lifestyle [16]. The second messenger c-di-GMP acts through an alternative regulation pathway and seems to promote biofilm development by a variety of routes: repression of motility-related genes, exopolysaccharide overproduction and expression of the adhesin CdrA [17, 18]. RsmA and c-di-GMP share overlapping targets and indirectly regulate each other with antagonistic effects, supporting the notion of a redundant system [19]. The fact that the c-di-GMP positively regulated efflux pump overexpression through *brlR* induction highlights the importance of the messenger in the development of antibiotic resistance/persistence phenotypes [20].

In sessile cells, the action of antibiotics is contrasted by a variety of mechanisms that make them less susceptible to antimicrobials than planktonic cells [21]. Notably, the biofilm matrix acts as a barrier, limiting the diffusion of toxic compounds [22]; in particular, binding to eDNA prevents positively charged antibiotics such as aminoglycosides from penetrating the bacterial cells [23]. Moreover, in biofilm-growing *P. aeruginosa* a wide range of resistance determinants are expressed or upregulated in a biofilm-specific manner [24]. Indeed, overexpression of the efflux pumps – particularly MexAB-OprM and MexXY-OprM – is the main cause of the multiple antibiotic-resistant phenotype [25] that characterizes chronic *P. aeruginosa* infection and contributes to the failure of its eradication in CF patients [26, 27]. The *mexAB-oprM* operon is upregulated in biofilms resistant to azithromycin [28] and fluoroquinolones [29] and also seems to be involved in colistin tolerance, which has been described in actively growing *P. aeruginosa* cells [30]. MexXY-OprM is the main aminoglycoside resistance determinant. It is a typical example of inducible adaptive resistance [31]; this is also demonstrated by the frequent recovery, from chronic patients, of strains bearing mutations in *mexZ*, a repressor gene of the *mex-XY* operon, which is considered as a mutation hotspot in biofilm-growing *P. aeruginosa* and a typical example of convergent evolution of different CF clonal lineages [32, 33]. Other remarkable examples of antibiotic resistance associated with biofilm growth are endogenous AmpC ß-lactamase overexpression and upregulation of the *ndvB* gene [34]; the latter is involved in biofilm-specific synthesis of cyclic glucans, which are responsible for aminoglycoside binding and trapping [35]. Finally, the biofilm is an ideal environment for HGT events [8], which contribute to the spread of resistance determinants. Conjugation events are favored by close contact between cells of different strains and/or species [36]; moreover, it has recently been suggested that *P. aeruginosa* biofilms can achieve a natural competence to acquire both genomic and plasmid DNA [37]. This is a cause of particular concern for chronic CF patients, whose lungs are often colonized by different antibiotic-resistant strains, a condition that has the potential to give rise to multidrug resistance [38].

2.2 *P. aeruginosa* CF lung infection

Cystic Fibrosis is a genetic autosomic disease due to mutations in the cystic fibrosis transmembrane conductance regulator (*CFTR*) gene, which involve a wide range of dysfunctions that alter the airway environment and increase susceptibility to bacterial respiratory infections. *CFTR* gene dysfunction affects epithelial cells, the pancreas (malabsorption), the liver (biliary cirrhosis), the sweat glands (heat shock) and the vas deferens (infertility). Patients with late disease suffer from bronchiectasis, small airway obstruction and progressive respiratory impairment [39]. CF is characterized by recurrent pulmonary exacerbations. Worsening of the chronic lung infection symptoms (particularly cough and sputum production), increased bacterial load and inflammation and, often, a reduction in FEV1 (forced respiratory volume in 1 second) impair lung function hence quality of life and overall survival.

The identification of effective treatments requires a greater understanding of the factors underpinning the exacerbations. Notably, the lung of CF patients is initially colonized by *Haemophilus influenzae* and *Staphylococcus aureus*; then, patients gradually become susceptible to infection with a variety of environmental Gram-negative bacteria carrying a broad range of constitutive and acquired antibiotic resistance determinants [39]. *P. aeruginosa* is the main pathogen triggering airway inflammation and the leading cause of CF morbidity and mortality [40]. Most CF patients are susceptible to *P. aeruginosa* respiratory infections from infancy. The 30% of them acquire a strain from the environment resulting in acute infections in the first year of life, this rate increases to about 50% before turning 3 years, while mucoid phenotype and chronic infection usually raise from 3 to 16 years [10].

Lung colonization generally involves alternate asymptomatic periods and relapses with progressive tissue deterioration that eventually lead to lung failure and to premature death. Over the years *P. aeruginosa* develops multiple phenotypic variants such as SCVs, mucoid and persistent forms. In particular, SCVs are typically isolated from the lungs of chronic CF patients. They are small (1–3 mm in diameter) usually non-motile and resistant to several classes of antibiotics; produce high amounts of exopolysaccharide and form biofilms that strongly adhere to surfaces [41]. *In vitro* and *in vivo* tests have demonstrated that exposure to sublethal concentrations of antibiotics, such as aminoglycosides, selects for SCVs. In CF patients, prolonged persistent infection, deterioration of pulmonary function and increased antibiotic resistance all correlate with SCVs detection in sputum [41].

P. aeruginosa adaptation to the CF lung environment ultimately results in a mucoid phenotype, a conversion first described by Lam and colleagues [42], which may take several months to years. The mucoid material has subsequently been identified as alginate. In mucoid strains, alginate may favor adhesion to lung epithelial cells, thereby inhibiting clearance. Nutrient restriction, dehydration and suboptimal antibiotic concentrations may result in mucoidity [7, 43]. Host inflammation responses are also believed to contribute to mucoid conversion, a hypothesis that is supported by the absence of mucoid variants among environmental isolates [44].

3. Persister development: antibiotic failure and microbiological diagnosis

3.1 Persistent and viable but non-culturable (VBNC) bacterial forms

P. aeruginosa lung infections tend to be recurrent. Relapses are chiefly due to the development of persisters, bacterial forms that are unsusceptible to antibiotics and often difficult to detect by routine microbiological assays. Persistence has been defined as "the ability of a subset of the bacterial population to survive to a bactericidal antibiotic concentration" [45]. Survival is demonstrated by bacterial growth in culture once the stressor, i.e. antibiotic concentrations several times higher than the minimal inhibitory concentration (MIC), has been removed and nutrients have been restored. Accordingly, the main features distinguishing persisters from resistant cells are the inability of the former cells to grow in presence of antibiotics, though viable and metabolically active, and the lack of heritability [45].

Persisters have been considered as dormant cells that are unaffected by antibiotics [46]. However, lack of significant growth or metabolic activity does not equal persistence, since the majority (> 99%) of dormant subpopulations are not true persisters. Persistence is a far more complex condition than dormancy [47], it shows an intense metabolic activity despite cell failure to grow or divide. Indeed, starvation-induced persisters produced more ATP per mol of carbon source consumed than nutrient supplied cells did [48]. Accordingly, persister cells seem to be able to catabolize carbon sources, which results in increased electron transport chain activity and membrane potential and increased aminoglycoside uptake [49]. However, although bacterial metabolic processes and persistence are closely related, the mechanisms involved are largely unclear [50].

Antibiotic persistence is not to be confused with antibiotic tolerance. In particular, whereas tolerance involves the whole bacterial population, persistence regards only a subset of specialized cells. Moreover, tolerant cells are killed, even if more slowly than susceptible cells, by high antibiotic doses while persisters are maintained over time (**Figure 1**). Notably, however, the two cell types share the same MIC as susceptible cells [45].

Two types of persisters have been described to date: stochastic and triggered. The former cells constitute a small subpopulation that can be found in all bacterial cultures, even in exponentially growing ones, whereas the latter are induced by environmental as well as host-related stressors. Unfavorable environmental conditions, e.g. nutrient and oxygen depletion, catabolite accumulation and suboptimal pH, which can induce persistence, can occur in the lungs of CF patients, especially in the deepest layers of *P. aeruginosa* biofilms [51]. Repeated antibiotic treatment directed at eradicating chronic infection can contribute to the induction of these specialized bacterial forms [26].

VBNC cells are dormant forms described in several bacterial species, including *P. aeruginosa*. They are characterized by the inability to grow on bacteriological media despite the presence of metabolic activity [52]. VBNC cells share several features with persisters, including a number of inducing factors of which the most common are starvation, oxidative stress, suboptimal salinity and pH and low temperature [52]. Moreover, both phenotypes are highly resilient to antimicrobials. These similarities have led some researchers to conclude that "persister and VBNC cells actually represent subsequent stages of the same cycle of dormancy, adopted by non-sporulating bacteria to survive unfavorable conditions" [52]. According to this theory, stress exposure would induce the development of persisters, which in case of prolonged exposure would turn into VBNC cells, whereas stressor removal

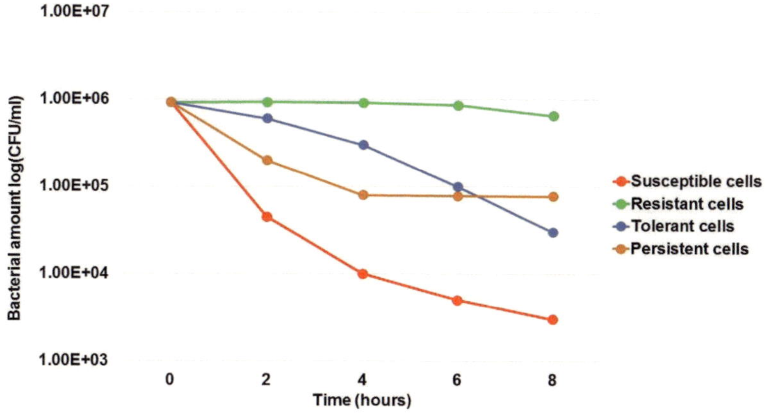

Figure 1.
Behavior of susceptible, resistant, tolerant and persistent bacterial subpopulations treated with antibiotic concentrations exceeding the MIC. CFU: Colony forming unit.

and nutrient restoration would involve recovery of the full metabolic state typical of exponential growth [52, 53]. Unlike culturable persisters, VBNC cells can regain culturability only through the action of specific activators (**Figure 2**), a phenomenon known as resuscitation [54]. The activators can be specific for the bacterial species and even for a single strain; while not completely understood they seem easily found *in vivo* [54].

3.2 VBNC *P. aeruginosa* and issues related to the diagnosis of CF lung infection

In CF patients, the microbiological diagnosis of *P. aeruginosa* lung infection is still performed by culture-based assays, which cannot detect VBNC cells [6]. Such assays involve microorganism isolation using rich (Columbia blood or chocolate) and selective (MacConkey or Pseudomonas) agar followed by isolate identification by biochemical or mass spectrometry analysis [55, 56]. Though effective in diagnosing primary colonization and pulmonary exacerbations, these methods suffer from considerable limitations, first and foremost poor sensitivity, due to the multiple phenotypic variants found in *P. aeruginosa* isolated from chronic CF biofilm-related infections [57].

A variety of stressors, principally nutrient depletion, oxidative and osmotic stress, an acid pH, a strong immune response and the presence of subinhibitory antibiotic concentrations [51, 58], make the CF lung an unfavorable environment for *P. aeruginosa*. The bacterial response involves the development of different phenotypes. The best known is the mucoid phenotype [59], alongside the loss of motility and pigmentation [60], the formation of auxotrophic variants and SCVs [61]. All these phenotypes are characterized by slow growth, which hampers culture-based diagnosis. However, the main problem is detecting VBNC cells. These cells – albeit not necessarily virulent – given suitable conditions can revert to full metabolically active forms capable of quick duplication and full virulence [54], which trigger a new infection. Developing a diagnostic technique that detects these forms is therefore critical to forecast symptom relapse and start early treatment.

3.3 The multifaceted role of antibiotics

Antibiotic treatment can play two different roles as regards persistent cells: it can either select a pre-existing persistent subpopulation or induce the persistent phenotype [45]. The hypothesis has also been advanced that antibiotics exert a biphasic dose-dependent action, i.e. inhibition of bacterial growth at high (≥ the MIC) doses

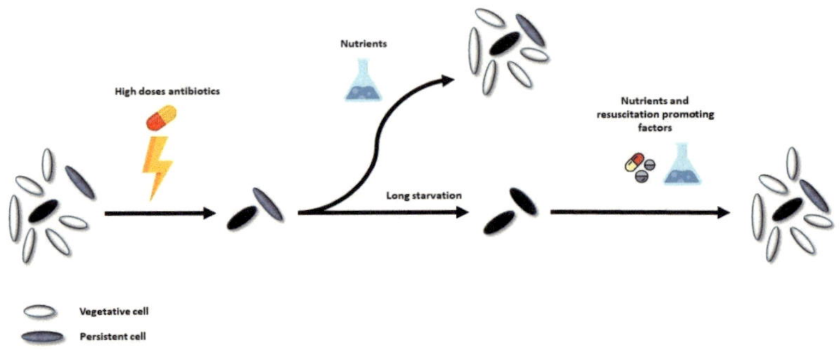

Figure 2.
Differences between culturable persistent and VBNC cells after stressor removal and nutrient restoration. Whereas persistent cells quickly begin to grow and divide again, VBNC cells require exposure to a growth activator (the resuscitation-promoting factor) before regaining full metabolic activity and doubling ability. The progeny of both cell types will contain a mixed population as the starting culture.

and stimulation of a specific bacterial response by acting as a molecular signal at low concentrations (< the MIC), a phenomenon known as hormesis [62].

Failed infection eradication even after prolonged antibiotic treatment is a major clinical problem in patients with microbial biofilms. Antibiotic unresponsiveness has been explained by poor drug penetration in the biofilm matrix and by the development of dormant/persistent cells in the deepest biofilm layers [22]. Since low antibiotic concentrations are probably found for extended periods in the lung of CF patients with intermittent/chronic infection, who undergo repeated antibiotic treatment, the development of persistent forms is likely to be stimulated by the drugs themselves. Evidence to this effect has been reported for different classes of antibiotics, including quinolones and aminoglycosides, although more comprehensive investigations are required to draw firm conclusions.

4. *P. aeruginosa* VBNC cell detection and quantification in CF respiratory samples

4.1 Total live cell detection strategies

Given the wide phenotypic variability of *P. aeruginosa*, encompassing difficult-to grow phenotypes, several culture-independent approaches have been devised to provide reliable infection diagnosis. DNA-based techniques are useful because they are able to detect the whole bacterial population. Most protocols are based on PCR or qPCR assays [63].

To find a suitable target gene on which to base *P. aeruginosa* detection, most protocols have been tested on a variety of bacterial isolates of different origins. The *oprL* gene, encoding a peptidoglycan-associated protein, has long been considered as one of the best targets [64–67]. However, its specificity was questioned when Anuj and colleagues [68] obtained cross-reactions with other species. Notably, the selection of multiple targets is considered as the best approach to *P. aeruginosa* detection, since it excludes false-negative results due to mutations in the amplified gene sequences. The *gyrB* and *ecfX* genes are two other widely used targets. The former gene encodes the DNA gyrase subunit B. Tests against several CF *P. aeruginosa* isolates have identified a species-specific internal sequence [69, 70]. The *ecfX* gene – found in 19 copies/genome – encodes a σ factor belonging to the ECF subfamily, which is involved in the synthesis of proteins with an extracytoplasmic function and seems to play a role in *P. aeruginosa* haem uptake and virulence [71]. The gene has been reported to be specific of *P. aeruginosa* and has been used to achieve its detection in environmental as well as clinical samples [6, 71]. Further proposed targets are the *algD* gene [72] and some 16 s [73] and 23 s rDNA sequences [74].

After reliable detection, a key issue is direct pathogen quantification in sputum samples. Most of the work in this field has been performed after 2010 using specific extraction kits and lysis protocols [66, 67, 74–76].

The main drawback of DNA-based approaches is that they do not detect only live bacterial cells and may be affected by the presence of dead cells as well as by eDNA [77]. An efficient and widely used approach to this problem is to treat samples with propidium monoazide before DNA extraction [66]. The dye penetrates the cells via damaged walls/membranes and binds their DNA; after photo-activation, binding to the nucleic acids prevents DNA polymerase binding, hence DNA amplification in PCR assays. Since live cells commonly have an intact wall, they are not affected by the dye and only their DNA is detected. The same objective can be achieved with other treatments such as ethidium monoazide and DNase [6, 77].

Despite some drawbacks, DNA-based methods provide additional valuable information to the cell culture results when investigating and monitoring *P. aeruginosa* colonization dynamics in the CF lung [78]. Accordingly, extensive metagenomic studies of the CF microbiota have highlighted that persistent cells play a major role in infection chronicization and that persistence is favored by alterations in bacterial gene expression [79], further stressing the value of molecular techniques in routine diagnostics [80, 81].

Another useful technique capable of providing direct bacterial quantification is flow cytometry. Although it has mostly been employed to investigate bacterial physiology and metabolic responses [82], efforts to optimize its quantification ability have made it suitable for some diagnostic applications [83]. In particular, flow cytometry analysis and imaging now enable detection and enumeration of non-culturable and intracellular *P. aeruginosa* cells [84, 85].

Other approaches to detect the whole microbial community of CF lung have also been developed and in the last years indirect detection has been also achieved by metabolomic methods targeting specific bacterial metabolites as pathogen footprints [86].

4.2 Evidence of the presence of VBNC *P. aeruginosa* in CF sputum

The presence of VBNC *P. aeruginosa* cells in the CF lung and in particular their role in infection recurrence are highly controversial. However, the induction of VBNC cells in the CF lung environment currently seems to be the most likely explanation for the failure of infection eradication in the presence of a negative microbiological diagnosis [6].

The first reports of pathogen persistence in patients with negative sputum cultures, published by Schelstraete and Deschaght and colleagues [87], described the swift reappearance of the same *P. aeruginosa* strain, after a brief interval of ostensible resolution, in patients treated with eradication therapy. Deschaght and co-workers [66] subsequently demonstrated that the pathogen could be detected by qPCR much earlier than by culture assays and that qPCR was able to detect a high percentage (62%) of non-culturable *P. aeruginosa* cells in sputum samples from patients who had received the first week of antibiotic treatment. A discrepancy between culture-based and culture-independent methods has also been reported by Le Gall [75] and Héry-Arnaud [76] who advanced the hypothesis of a shift of bacterial cells to a non-culturable state. A positive qPCR assay preceding a positive culture has also been described by McCulloch and colleagues [74] and, more recently, by Boutin and co-workers [88].

Our group has carried out extensive work to identify and quantify VBNC *P. aeruginosa* cells in CF sputum [6]. Combining two previously published *ecfX*-targeting primers we obtained a new amplicon (145 bp) suitable for qPCR. Testing of the new primer pair against a panel of 115 *P. aeruginosa* strains of different origins and other Gram-negative bacterial species failed to elicit a cross-reaction, confirming the species specificity of the selected target. Moreover – even though the use of a single target gene cannot exclude false-negative results due to target mutations [68] – the *ecfX* sequence yielded a positive PCR result in 111/115 (96.6%) of the *P. aeruginosa* strains and the use of a second target gene (*gyrB*) did not lead to an increase of *P. aeruginosa* detection ([89] unpublished data).

Total DNA was extracted from CF sputum samples using the QIAamp DNA kit (Qiagen, Hilden, Germany) and qPCR assays were performed using a SYBR Green reaction format. The sensitivity of the protocol combining DNA extraction and qPCR was determined by testing *P. aeruginosa*-free sputum samples inoculated with serial dilutions of log phase *P. aeruginosa* cultures. Protocol sensitivity was 70 cells/ml, which is comparable to the sensitivity of TaqMan probe-based qPCR assays [90]. Its limit of detection, determined by amplifying serial dilutions of a purified *ecfX* amplicon, was 5.2 x 10^{-9} ng/reaction, corresponding to about 140 cells/ml in the original samples.

eDNA interference was excluded by treating samples with DNase I (18 U) before DNA extraction. Preliminary assays were performed using *P. aeruginosa*-free sputum samples inoculated with 10% live and 90% dead *P. aeruginosa* cultures. DNase-treated and untreated aliquots were processed using an in-house crude extraction procedure or the QIAamp extraction kit (**Figure 3**).

The qPCR counts of DNase-treated aliquots always matched the live cell quota (10%) of the *P. aeruginosa* inoculum. As regards the untreated aliquots, they corresponded to the whole *P. aeruginosa* load (100%) when qPCR was performed on crude extracts, whereas qPCR performed on DNA extracted with the QIAamp kit yielded counts that were comparable to those obtained after DNase pretreatment. This is likely due to the fact that the eDNA of dead *P. aeruginosa* cells was too damaged to be efficiently bound and retained in the extraction column. It can thus be assumed that DNA extraction with suitable commercial kits – whether alone or combined with DNase treatment – excluded eDNA and provided reliable quantification of live bacterial cells.

We performed the same procedure in 88 CF sputum samples from 55 patients.

The qPCR and culture-based counts were largely comparable (i.e. 78.41% of all samples, 43.18% culture-negative and 35.23% culture-positive). Notably, the absence of samples that were simultaneously culture-positive and qPCR-negative excluded false negatives. The most interesting results were those where the qPCR count exceeded the culture-based count (11.40% of samples) and those where culture-negative samples showed a qPCR-positive result (10.23%). Given eDNA exclusion by DNase treatment and DNA extraction procedure, the discrepancy was held to reflect the presence of VBNC *P. aeruginosa* cells, in line with data reported by Deschaght [66], Le Gall [75], McCulloch [74] and Boutin [88]. Crucially, 1 and 3 months after the PCR-positive results, the cultures turned positive in 2 patients.

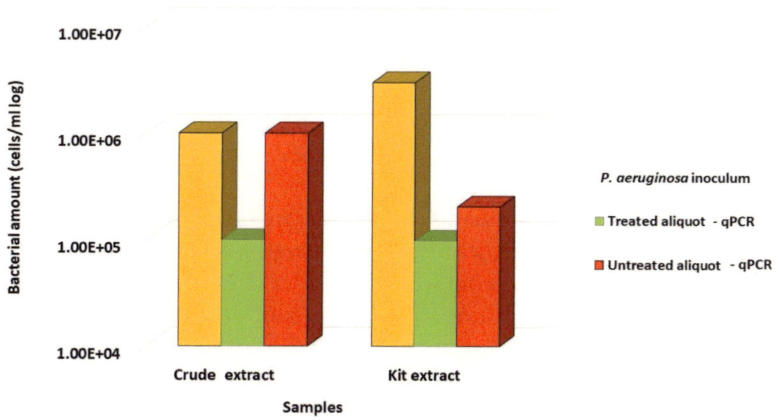

Figure 3.
(Modified from Ref. [6]) Detection of live P. aeruginosa *cells in DNase-treated sputum samples.* P. aeruginosa *abundance was quantified by qPCR in CF sputum samples inoculated with 10% live and 90% dead cultures with/without DNase I pretreatment. DNA was extracted with a crude extraction procedure or with the QIAlamp extraction kit. The qPCR counts were compared to the whole bacterial inoculum.*

5. Evaluation of the possible role of antibiotics and other stressors in triggering VBNC cells and culturable persisters in *P. aeruginosa* biofilms

After demonstrating VBNC *P. aeruginosa* cells in the lungs of CF patients (6), a key issue was to establish the factors that trigger their induction. We therefore examined the role of antibiotic treatments and of some environmental stressors that are found in the CF lung [51, 58], in selecting and maintaining persister cells, particularly VBNC cells.

5.1 Evaluation of the role of sub-MIC of ciprofloxacin and tobramycin in inducing VBNC and culturable persistent cells in starved *P. aeruginosa in vitro* biofilms

The possibility of VBNC cell induction by low antibiotic doses was first explored by our group in starved *S. aureus* biofilms. Greater VBNC cell persistence and survival were found in starved cultures exposed to vancomycin and quinupristin/dalfopristin than in those subjected to starvation alone [91].

Similar experiments were subsequently performed to test VBNC cell induction in *P. aeruginosa* biofilms exposed to starvation, alone or combined with subinhibitory (1/4 x MIC) concentrations of ciprofloxacin or tobramycin [92], which are the most widely used treatments for *P. aeruginosa* lung infection [39, 93]. Biofilms of the laboratory strain *P. aeruginosa* PAO1 and the CF isolate *P. aeruginosa* C24 were developed in rich medium in 35 mm petri dishes for 48 h at 37°C and then subcultured in non-nutrient (NN) broth, alone or supplemented with sub-MIC concentrations of ciprofloxacin or tobramycin for 170 days. The medium was refreshed once a week. Samples were evaluated at 60, 75, 90, 120, 135, 150 and 170 days. The biofilm content in VBNC cells was determined as follows:

- the culturable population was quantified by plate counts performed on cystine lactose electrolyte-deficient (CLED) agar after incubation for 24, 28 and 72 h at 37°C;

- total viable cells (TVCs) were expressed as the average of the counts obtained from *ecfX*-qPCR and flow cytometry assays after live/dead staining using SYBR Green 1x and propidium iodide 40 μg/ml;

- the number of VBNC cells was determined as the difference between TVCs and culturable cells (only differences ≥0.5 log were considered).

The results are reported in **Figure 4** and are expressed as percentage of TVCs.

Unlike the *S. aureus* biofilms, a culturable *P. aeruginosa* subpopulation, which can be defined as triggered persisters, was detected throughout the experiment. A VBNC subpopulation also developed and was more abundant in presence of the antibiotics. In particular, a subset of VBNC *P. aeruginosa* PAO1 cells was detected in all conditions and gradually declined in starved and ciprofloxacin-exposed biofilms; in contrast, the VBNC subpopulation triggered by sub-MIC tobramycin exceeded 90% of TVCs until the end of the experiment. In *P. aeruginosa* C24, starvation alone induced a discontinuous VBNC subpopulation; starvation and ciprofloxacin triggered a persister population only between 75 and 135 days; and starvation and tobramycin induced a constant VBNC population whose abundance was similar to the one determined in the PAO1 strain at the end of the experiment.

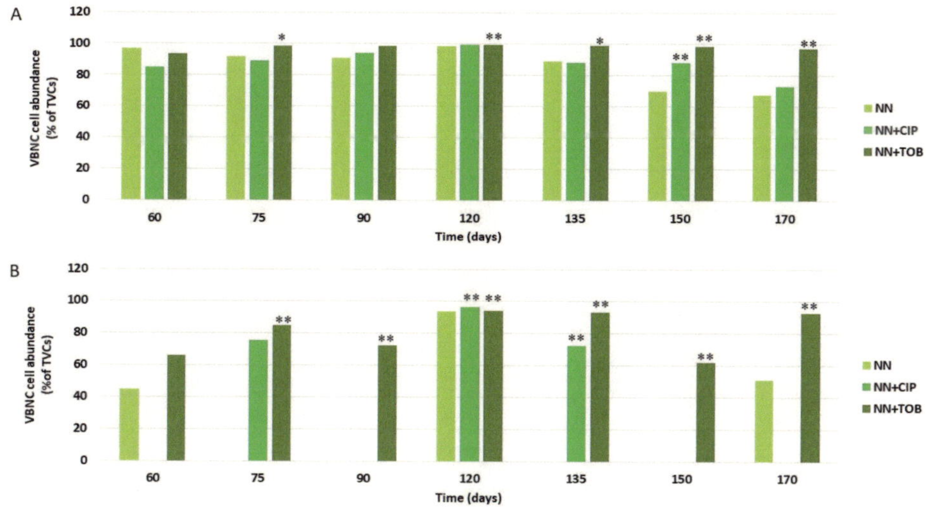

Figure 4.
(Modified from Ref. [92]). VBNC P. aeruginosa *cell abundance in starved biofilms exposed to sub-MIC antibiotic concentrations. The VBNC cell subpopulation was assessed at specific time points in* P. aeruginosa *PAO1 (A) and C24 (B) biofilms exposed to starvation (NN) or starvation+ ¼ x MIC ciprofloxacin (NN + CIP) or tobramycin (NN + TOB). VBNC cells were the difference between total viable cells (TVCs) and plate counts (only differences ≥0.5 log were considered). Cell abundance was compared in the three stress conditions.* $* = p < 0.05$, $** = p < 0.001$.

These data suggest that sub-MIC concentrations of tobramycin (not ciprofloxacin) play a strong role in inducing VBNC *P. aeruginosa* and are in line with previous reports of the contribution of protein synthesis inhibitors [94] and aminoglycosides [95] to VBNC cell induction. The demonstration of the role of toxin-antitoxin modules [96] and ribosome hibernation [3] in persisters development, exerted via a reduction of protein synthesis, further supports the observed behavior of tobramycin.

After documenting the role of subinhibitory drug concentrations in VBNC cell induction and maintenance, we examined the effectiveness of high antibiotic concentrations on *P. aeruginosa* biofilm eradication by evaluating the abundance of persistent and VBNC cells in mature *P. aeruginosa* PAO1 biofilms exposed to 1000 x MIC/100 x MBEC (minimal biofilm eradication concentration) tobramycin for 24 h (**Figure 5**).

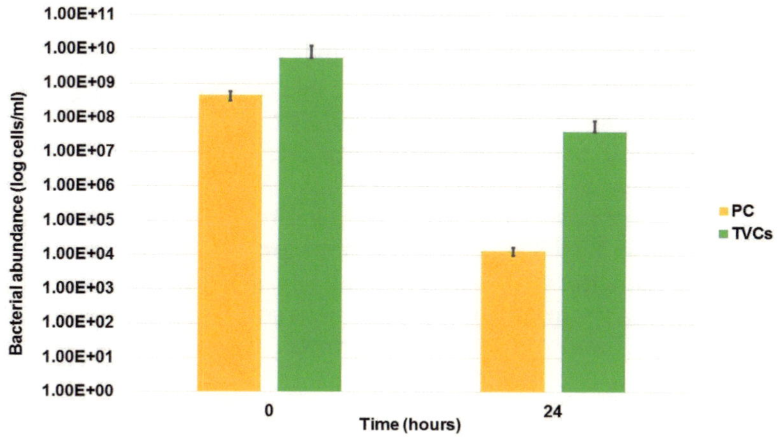

Figure 5.
Biofilm persistence to tobramycin treatment. 24-hour-old P. aeruginosa *PAO1 biofilms were exposed for 24 h to tobramycin 1000 x MIC and assessed for their content in culturable persisters and VBNC cells before and after antibiotic treatment. Persisters were determined by plate count (PC), whereas total viable cells (TVCs) were determined by* ecfX*-qPCR and live/dead flow cytometry. The VBNC population was the difference between TVCs and PCs. The results are given as the average of three biological replicates ± standard deviation.*

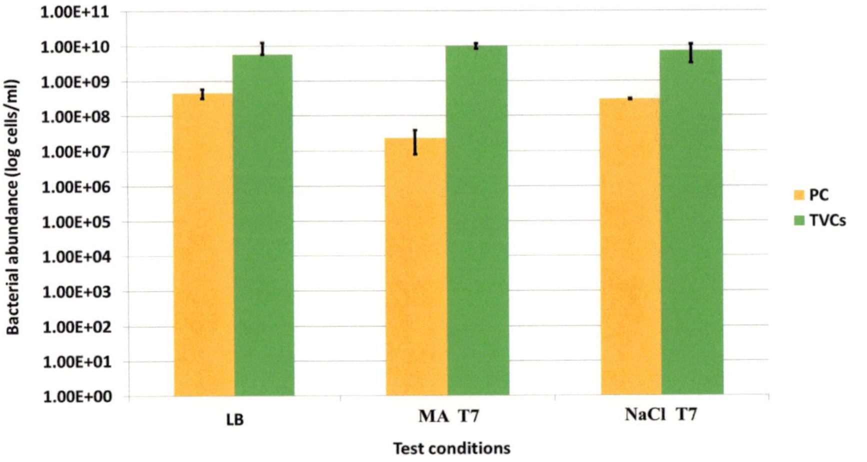

Figure 6.
Induction of VBNC P. aeruginosa *cells in biofilms exposed to environmental stress factors.* P. aeruginosa *PAO1 biofilms were grown in Luria Bertani broth, with metabolite accumulation (MA T7) or in LB + 13 g/l NaCl for 7 days (NaCl T7). Culturable cells were determined by plate count (PC) and total viable cells (TVCs) were determined by qPCR/flow cytometry assays. VBNC cells were the difference between TVCs and culturable cells. These values were compared to those determined in a 24-hour-old biofilm grown in Luria Bertani (LB) broth. The results are average of three biological replicates ± standard deviation.*

As expected, the high tobramycin concentrations chiefly affected the culturable population, which showed a reduction >4 log, whereas the TVC counts showed a 2 log reduction, highlighting the presence of more than 1 x 10^7 VBNC cells/ml after 24-hours exposure to 1000 x MIC tobramycin.

5.2 Evaluation of the possible involvement of additional stressors found in the CF lung

Finally, we examined the possible contribution of further environmental factors – especially the high salinity and metabolite accumulation that are found in the CF lung [51, 58], – in the induction of persistent and VBNC *P. aeruginosa* cells.

Culturable cells and TVCs were counted as described above in *P. aeruginosa* PAO1 biofilms grown for 7 days in Luria Bertani (LB) broth, alone or added with 13 g/l NaCl. The counts were compared to those of a mature 24-hour-old biofilm (**Figure 6**).

As shown in the diagram, *P. aeruginosa* biofilms tolerate and adapt to the high salinity found in the CF lung, since exposure to this stressor for 7 days failed to induce a significant difference in culturable and VBNC cell amount compared to the control condition. In contrast, the plate counts demonstrated a difference of 1 log between 7-day-old and 24-hour-old LB biofilms, whereas the number of TVCs at the two time points was not significantly different. This indicates a shift of *P. aeruginosa* cells to the VBNC state in biofilms grown for 7 days in LB medium, where bacterial metabolites accumulate. Most likely, nutrient reduction and waste accumulation induce a major shift to the persistent state, as also demonstrated for the VBNC cells in biofilms maintained in NN broth.

6. Conclusions

The generation of persistent cell subpopulations is a bacterial survival strategy against adverse environmental conditions [2]. Whereas stochastic persisters are rare, external stressors can convert most bacterial population into persistent cells [45].

In infectious biofilms, a combination of stress factors can induce the development of persistent forms which can trigger infection recurrence. This is a cause of special concern in *P. aeruginosa* CF lung infection, where VBNC cells undermine treatment and hamper microbiological diagnosis, which is still routinely performed by culture-based assays. A routine diagnostic workup including culture-independent approaches should thus be urgently adopted.

The evidence described in this chapter demonstrates that VBNC *P. aeruginosa* cells are found in sputum samples from CF patients and that several months after a culture-negative and qPCR-positive assay some patients experience infection recurrence and have culture-positive sputum samples. These data also highlight the reliability of qPCR in detecting the whole bacterial population, including the phenotypic variants that are missed by culture-based assays. Notably, flow cytometry has demonstrated the accuracy of the *ecfX*-targeting qPCR protocol in detecting all viable *P. aeruginosa* cells and can provide a sound alternative for routine monitoring of the infection. Together, qPCR and flow cytometry supply a clear picture of *P. aeruginosa* population dynamics in the lungs of CF patients with intermittent and chronic infection and can be harnessed to monitor the effectiveness of the antibiotic therapy and to foster the development of new eradication treatments.

The two techniques have enabled us to gain insight into the role of antibiotics in VBNC cell development and infection persistence. Notably, whereas high antibiotic doses can select persistent subpopulations, subinhibitory concentrations – which are found in the CF lung between treatment cycles and in the deepest biofilm layers [51, 58] – can stimulate the development of persistent phenotypic variants, including VBNC cells [53]. Besides the fact that starvation proved to be a necessary condition for VBNC cell induction in our *in vitro* biofilms, our findings highlight a different behavior of tobramycin and ciprofloxacin. Although large amounts of VBNC cells were induced in all test conditions over the first 120 days, their number was maintained more consistently in presence of tobramycin, whereas ciprofloxacin exerted a discontinuous effect similar to the one of starvation alone. These findings can partly be explained by the ability of tobramycin to act as a signal molecule that interferes with QS signals, thus modulating gene expression in biofilm-growing *P. aeruginosa* [58], and by its adverse effect on protein synthesis via ribosome binding [3].

In conclusion, more detailed information on the main gene pathways and persistence regulators and on the effects of different antibiotics is essential to meet the challenge of antibiotic-resilient *P. aeruginosa* infectious biofilms and the eradication of CF lung infection.

Acknowledgements

We are grateful to Dr. Salvatore Vaiasicca for his contribution to flow cytometry assays and to Dr. Natalia Cirilli for helpful discussion on *P. aeruginosa* CF lung infections.

This work was supported by FFC grants # 13/2017 and #16/2019 of the Italian Cystic Fibrosis Research Foundation.

Author details

Gianmarco Mangiaterra*, Mehdi Amiri, Nicholas Cedraro and Francesca Biavasco
Department of Life and Environmental Sciences, Polytechnic University of Marche, Ancona, Italy

*Address all correspondence to: g.mangiaterra@pm.univpm.it

References

[1] Costerton JW, Stewart PS, Greenberg EP: Bacterial biofilms: a common cause of persistent infections. Science. 1999; 284: 1318-22. DOI: 10.1126/science.284.5418.1318.

[2] Ayrapetyan M, Williams T, Oliver JD: Relationship between the Viable but Nonculturable State and Antibiotic Persister Cells. J Bacteriol. 2018; 200: e00249-18. DOI: 10.1128/JB.00249-18.

[3] Wood TK, Song S, Yamasaki R: Ribosome dependence of persister cell formation and resuscitation. J Microbiol. 2019; 57: 213-219. DOI: 10.1007/s12275-019-8629-2.

[4] Stewart PS and Bjarnsholt T: Risk factors for chronic biofilm-related infection associated with implanted medical devices. Clin Microbiol Infect. 2020; 26: 1034-1038. DOI: 10.1016/j.cmi.2020.02.027.

[5] Pasquaroli S, Zandri G, Vignaroli C, Vuotto C, Donelli G, Biavasco F: Antibiotic pressure can induce the viable but non-culturable state in *Staphylococcus aureus* growing in biofilms. J Antimicrob Chemother. 2013; 68: 1812-1817. DOI: 10.1093/jac/dkt086.

[6] Mangiaterra G, Amiri M, Di Cesare A, Pasquaroli S, Manso E, Cirilli N, Citterio B, Vignaroli C, Biavasco F: Detection of viable but non-culturable *Pseudomonas aeruginosa* in cystic fibrosis by qPCR: a validation study. BMC Infect Dis. 2018; 18: 701. DOI: 10.1186/s12879-018-3612-9.

[7] Donlan RM and Costerton JW: Biofilms: survival mechanisms of clinically relevant microorganisms. Clin Microbiol Rev. 2002; 15: 167-193. DOI: 10.1128/cmr.15.2.167-193.2002.

[8] Hall CW and Mah TF: Molecular mechanisms of biofilm-based antibiotic resistance and tolerance in pathogenic bacteria. FEMS Microbiol Rev. 2017; 41: 276-301. DOI: 10.1093/femsre/fux010.

[9] Azimi S, Klementiev AD, Whiteley M, Diggle SP: Bacterial Quorum Sensing During Infection. Annu Rev Microbiol. 2020; 74: 201-219. DOI: 10.1146/annurev-micro-032020-093845.

[10] Moradali MF, Ghods S, Rehm BHA: *Pseudomonas aeruginosa* Lifestyle: A Paradigm for Adaptation, Survival, and Persistence. Front Cell Infect Microbiol. 2017; 7: 39. DOI: 10.3389/fcimb.2017.00039.

[11] Olsen I: Biofilm-specific antibiotic tolerance and resistance. Eur J Clin Microbiol Infect Dis. 2015; 34: 877-886. DOI: 10.1007/s10096-015-2323-z.

[12] Flemming HC and Wingender J: The biofilm matrix. Nat Rev Microbiol. 2010; 8: 623-633. DOI: 10.1038/nrmicro2415.

[13] Valentini M, Gonzalez D, Mavridou DA, Filloux A: Lifestyle transitions and adaptive pathogenesis of *Pseudomonas aeruginosa*. Curr Opin Microbiol. 2018; 41:15-20. DOI: 10.1016/j.mib.2017.11.006.

[14] Rojo-Molinero E, Macià MD, Oliver A. Social Behavior of Antibiotic Resistant Mutants Within *Pseudomonas aeruginosa* Biofilm Communities. Front Microbiol. 2019; 10: 570. DOI: 10.3389/fmicb.2019.00570.

[15] Lee K and Yoon SS: *Pseudomonas aeruginosa* Biofilm, a Programmed Bacterial Life for Fitness. J Microbiol Biotechnol. 2017; 27: 1053-1064. DOI: 10.4014/jmb.1611.11056.

[16] Brencic A and Lory S: Determination of the regulon and identification of novel mRNA targets of *Pseudomonas aeruginosa* RsmA. Mol Microbiol. 2009; 72: 612-632. DOI: 10.1111/j.1365-2958.2009.06670.x.

[17] Colley B, Dederer V, Carnell M, Kjelleberg S, Rice SA, Klebensberger J: SiaA/D Interconnects c-di-GMP and RsmA Signaling to Coordinate Cellular Aggregation of *Pseudomonas aeruginosa* in Response to Environmental Conditions. Front Microbiol. 2016; 26: 7:179. DOI: 10.3389/fmicb.2016.00179.

[18] Valentini M and Filloux A: Biofilms and Cyclic di-GMP (c-di-GMP) Signaling: Lessons from *Pseudomonas aeruginosa* and Other Bacteria. J Biol Chem. 2016; 291: 12547-12555. DOI: 10.1074/jbc.R115.711507.

[19] Chen Y, Yuan M, Mohanty A, Yam JK, Liu Y, Chua SL, Nielsen TE, Tolker-Nielsen T, Givskov M, Cao B, Yang L. Multiple diguanylate cyclase-coordinated regulation of pyoverdine synthesis in *Pseudomonas aeruginosa*. Environ Microbiol Rep. 2015; 7: 498-507. DOI: 10.1111/1758-2229.12278.

[20] Liao J and Sauer K: The MerR-like transcriptional regulator BrlR contributes to *Pseudomonas aeruginosa* biofilm tolerance. J. Bacteriol. 2012; 194: 4823-4836. DOI: 10.1128/JB.00765-12.

[21] Penesyan A, Gillings M, Paulsen IT: Antibiotic discovery: combatting bacterial resistance in cells and in biofilm communities. Molecules. 2015; 20: 5286-5298. DOI: 10.3390/molecules20045286.

[22] Olivares E, Badel-Berchoux S, Provot C, Prévost G, Bernardi T, Jehlet F: Clinical Impact of Antibiotics for the Treatment of *Pseudomonas aeruginosa* Biofilm Infections. Front. Microbiol. 2020; 9: 399. DOI: 10.3389/fmicb.2019.02894.

[23] Wilton M, Charron-Mazenod L, Moore R, Lewenza S: Extracellular DNA Acidifies Biofilms and Induces Aminoglycoside Resistance in *Pseudomonas aeruginosa*. Antimicrob Agents Chemother. 2015; 60: 544-553. DOI: 10.1128/AAC.01650-15.

[24] Soares A, Alexandre K, Etienne M: Tolerance and Persistence of *Pseudomonas aeruginosa* in Biofilms Exposed to Antibiotics: Molecular Mechanisms, Antibiotic Strategies and Therapeutic Perspectives. Front Microbiol. 2020; 11: 2057. DOI: 10.3389/fmicb.2020.02057

[25] Soto SM: Role of efflux pumps in the antibiotic resistance of bacteria embedded in a biofilm. Virulence. 2013; 4: 223-229. DOI: 10.4161/viru.23724.

[26] Mulcahy LR, Burns JL, Lory S, Lewis K: Emergence of *Pseudomonas aeruginosa* strains producing high levels of persister cells in patients with cystic fibrosis. J Bacteriol. 2010; 192: 6191-6199. DOI: 10.1128/JB.01651-09.

[27] Horna G, López M, Guerra H, Saénz Y, Ruiz J: Interplay between MexAB-OprM and MexEF-OprN in clinical isolates of *Pseudomonas aeruginosa*. Sci Rep. 2018; 8: 16463. DOI: 10.1038/s41598-018-34694-z.

[28] Gillis RJ, White KG, Choi KH, Wagner WE, Schweizer HP, Iglewski BH: Molecular basis of azithromycin-resistant *Pseudomonas aeruginosa* biofilms. Antimicrob Agents Chemother. 2005; 49: 3858-3867. DOI: 10.1128/AAC.49.9.3858-3867.2005.

[29] Poole K: *Pseudomonas aeruginosa*: Resistance to the Max. Front Microbiol. 2011; 2: 65. DOI: 10.3389/fmicb.2011.00065.

[30] Pamp SJ, Gjermansen M, Johansen HK, TolkerNielsen T: Tolerance to the antimicrobial peptide colistin in *Pseudomonas aeruginosa* biofilms is linked to metabolically active cells, and depends on the *pmr* and *mexAB-oprM* genes. Mol Microbiol 2008; 68: 223-240. DOI: 10.1111/j.1365-2958.2008.06152.x.

[31] Morita Y, Tomida J, Kawamura Y: MexXY multidrug efflux system of *Pseudomonas aeruginosa*. Front

Microbiol. 2012; 3: 408. DOI: 10.3389/fmicb.2012.00408.

[32] Prickett MH, Hauser AR, McColley SA, Cullina J, Potter E, Powers C, Jain M: aminoglycoside resistance of *pseudomonas aeruginosa* in cystic fibrosis results from convergent evolution in the *mexz* gene. Thorax. 2017; 72:40-47. DOI: 10.1136/thoraxjnl-2015-208027.

[33] Colque CA, Albarracín Orio AG, Feliziani S, Marvig RL, Tobares AR, Johansen HK, Molin S, Smania AM: Hypermutator *Pseudomonas aeruginosa* Exploits Multiple Genetic Pathways To Develop Multidrug Resistance during Long-Term Infections in the Airways of Cystic Fibrosis Patients. Antimicrob Agents Chemother. 2020; 64: e02142-19. DOI: 10.1128/AAC.02142-19.

[34] Zhao J, Jiang H, Cheng W, Wu J, Zhao J, Wang J, Dong L: The role of quorum sensing system in antimicrobial induced *ampC* expression in *Pseudomonas aeruginosa* biofilm. J Basic Microbiol. 2015; 55: 671-678. DOI: 10.1002/jobm.201300987.

[35] Mah TF, Pitts B, Pellock B, Walker GC, Stewart PS, O'Toole GA: A genetic basis for *Pseudomonas aeruginosa* biofilm antibiotic resistance. Nature. 2003; 426: 306-310. DOI: 10.1038/nature02122.

[36] Molin S and Tolker-Nielsen T: Gene transfer occurs with enhanced efficiency in biofilms and induces enhanced stabilisation of the biofilm structure. Curr Opin Biotechnol. 2003; 14: 255-261. DOI: 10.1016/s0958-1669(03)00036-3.

[37] Nolan LM, Turnbull L, Katrib M, Osvath SR, Losa D, Lazenby JJ, Whitchurch CB: *Pseudomonas aeruginosa* is capable of natural transformation in biofilms. Microbiology. 2020; 166: 995-1003. DOI: 10.1099/mic.0.000956.

[38] Hughes D and Andersson DI: Environmental and genetic modulation of the phenotypic expression of antibiotic resistance. FEMS Microbiol Rev. 2017; 41: 374-391. DOI: 10.1093/femsre/fux004.

[39] Elborn JS: Cystic fibrosis. Lancet. 2016; 388: 2519-2531. DOI: 10.1016/S0140-6736(16)00576-6.

[40] Waters CM and Goldbergb JB: *Pseudomonas aeruginosa* in cystic fibrosis: A chronic cheater. Proc Natl Acad Sci U S A. 2019; 116: 6525-6527. DOI: 10.1073/pnas.1902734116.

[41] Malone JG: Role of small colony variants in persistence of *Pseudomonas aeruginosa* infections in cystic fibrosis lungs. Infect Drug Resist. 2015; 8: 237-247. DOI: 10.2147/IDR.S68214.

[42] Lam J, Chan R, Lam K, Costerton JW: Production of mucoid microcolonies by *Pseudomonas aeruginosa* within infected lungs in cystic fibrosis. Infect Immun. 1980; 28: 546-556.

[43] May TB, Shinabarger D, Maharaj R, Kato J, Chu L, DeVault JD, Roychoudhury S, Zielinski NA, Berry A, Rothmel RK, et al: Alginate synthesis by *Pseudomonas aeruginosa*: a key pathogenic factor in chronic pulmonary infections of cystic fibrosis patients. Clin Microbiol Rev. 1991; 4: 191-206. DOI: 10.1128/cmr.4.2.191.

[44] Schick A and Kassen R: Rapid diversification of *Pseudomonas aeruginosa* in cystic fibrosis lung-like conditions. Proc Natl Acad Sci U S A. 2018; 115: 10714-10719. DOI: 10.1073/pnas.1721270115.

[45] Balaban NQ, Helaine S, Lewis K, Ackermann M, Aldridge B, Andersson DI, Brynildsen MP, Bumann D, Camilli A, Collins JJ, Dehio C, Fortune S, Ghigo JM, Hardt WD, Harms A, Heinemann M, Hung DT, Jenal U, Levin BR, Michiels J, torz G, Tan MW, Tenson T, Van Melderen L, Zinkernagel

A: Definitions and guidelines for research on antibiotic persistence. Nat Rev Microbiol. 2019; 17: 441-448. DOI: 10.1038/s41579-019-0196-3.anderss.

[46] Wood TK, Knabel SJ, Kwan BW: Bacterial persister cell formation and dormancy. Appl Environ Microbiol. 2013; 79: 7116-7121. DOI:10.1128/AEM.02636-13.

[47] Orman MA and Brynildsen MP: Dormancy is not necessary or sufficient for bacterial persistence. Antimicrob Agents Chemother. 2013; 57: 3230-3239. DOI: 10.1128/AAC.00243-13.

[48] Radzikowski JL, Vedelaar S, Siegel D, Ortega ÁD, Schmidt A, Heinemann M: Bacterial persistence is an active sigmaS stress response to metabolic flux limitation. Mol Syst Biol. 2016; 12: 882. DOI: 10.15252/msb.20166998.

[49] Allison KR, Brynildsen MP, Collins JJ: Metabolite-enabled eradication of bacterial persisters by aminoglycosides. Nature. 2011; 473: 216-220. DOI: 10.1038/nature10069.

[50] Mohiuddin SG, Hoang T, Saba A, Karki P, Orman MA: Identifying Metabolic Inhibitors to Reduce Bacterial Persistence. Front Microbiol. 2020; 11: 472. DOI: 10.3389/fmicb.2020.00472.

[51] Bhagirath AY, Li Y, Somayajula D, Dadashi M, Badr S, Duan K: Cystic fibrosis lung environment and *Pseudomonas aeruginosa* infection. BMC Pulm Med. 2016; 16: 174. DOI: 10.1186/s12890-016-0339-5.

[52] Ayrapetyan M, Williams TC, Oliver JD: Bridging the gap between viable but non-culturable and antibiotic persistent bacteria. Trends Microbiol. 2015; 23: 7-13. DOI: 10.1016/j.tim.2014.09.004.

[53] Kim JS, Chowdhury N, Yamasaki R, Wood TK: Viable but non-culturable and persistence describe the same bacterial stress state. Environ Microbiol. 2018; 20: 2038-2048. DOI: 10.1111/1462-2920.14075.

[54] Li L, Mendis N, Trigui H, Oliver JD, Faucher SP: The importance of the viable but non-culturable state in human bacterial pathogens. Front Microbiol. 2014; 5: 258. DOI: 10.3389/fmicb.2014.00258.

[55] Gilligan P. Respiratory cultures from cystic fibrosis patients. In: Garcia LS, Isenberg HD, editors. Clinical microbiology procedures handbook. Washington DC: ASM press; 2007. p. 311-319.

[56] UK Cystic Fibrosis Trust Microbiology Laboratory Standards Working Group. Laboratory standards consensus document. Laboratory standards for processing microbiological samples from people with cystic fibrosis. (2010).

[57] Winstanley C, O'Brien S, Brockhurst MA. Pseudomonas aeruginosa Evolutionary Adaptation and Diversification in Cystic Fibrosis Chronic Lung Infections. Trends Microbiol. 2016; 24: 327-337. DOI: 10.1016/j.tim.2016.01.008.

[58] Andersson DI and Hughes D. Microbiological effects of sublethal levels of antibiotics. Nat Rev Microbiol. 2014; 12: 465-478. DOI: 10.1038/nrmicro3270.

[59] Malhotra S, Limoli DH, English AE, Parsek MR, Wozniak DJ. Mixed Communities of Mucoid and Nonmucoid *Pseudomonas aeruginosa* Exhibit Enhanced Resistance to Host Antimicrobials. mBio. 2018; 9:e00275-18. DOI: 10.1128/mBio.00275-18.

[60] Qin X, Zerr DM, McNutt MA, Berry JE, Burns JL, Kapur RP. Pseudomonas aeruginosa syntrophy in chronically colonized airways of cystic fibrosis patients. Antimicrob Agents

Chemother. 2012; 56: 5971-5981. DOI: 10.1128/AAC.01371-12.

[61] Lozano C, Azcona-Gutiérrez JM, Van Bambeke F, Sáenz Y: Great phenotypic and genetic variation among successive chronic Pseudomonas aeruginosa from a cystic fibrosis patient. PLoS One. 2018; 13:e0204167. DOI: 10.1371/journal.pone.0204167.

[62] Davies J, Spiegelman GB, Yim G: The world of subinhibitory antibiotic concentrations. Curr Opin Microbiol. 2006; 9: 445-453. DOI: 10.1016/j.mib.2006.08.006.

[63] Deschaght P, Van daele S, De Baets F, Vaneechoutte M: PCR and the detection of Pseudomonas aeruginosa in respiratory samples of CF patients. A literature review. J Cyst Fibros. 2011; 10: 293-297. DOI: 10.1016/j.jcf.2011.05.004.

[64] De Vos D, Lim Jr A, Pirnay JP, Struelens M, Vandenvelde C, Duinslaeger L, Vanderkelen A, Cornelis P: Direct detection and identification of *Pseudomonas aeruginosa* in clinical samples such as skin biopsy specimens and expectorations by multiplex PCR based on two outer membrane lipoprotein genes, *oprI* and *oprL*. J Clin Microbiol. 1997; 35: 1295-1299. DOI: 10.1128/JCM.35.6.1295-1299.1997.

[65] Xu J, Moore JE, Murphy PG, Millar BC, Elborn JS: Early detection of *Pseudomonas aeruginosa* – comparison of conventional versus molecular (PCR) detection directly from adult patients with cystic fibrosis (CF). Ann Clin Microbiol Antimicrob. 2004; 3: 21. DOI: 10.1186/1476-0711-3-21.

[66] Deschaght P, Schelstraete P, Van Simaey L, Vanderkercken M, Raman A, Mahieu L, Van Daele S, De Baets F, Vaneechoutte M. Is the improvement of CF patients, hospitalized for pulmonary exacerbation, correlated to a decrease in bacterial load? PLoS One. 2013; 8:e79010. DOI: 10.1371/journal.pone.0079010.

[67] BlanchardAC, Rooney AM, Yau Y, Zhang Y, Stapleton PJ, Horton E, Klingel M, Stanojevic S, Ratjen F, Coburn B, Waters V: Early detection using qPCR of *Pseudomonas aeruginosa* infection in children with cystic fibrosis undergoing eradication treatment. J Cyst Fibros. 2018; 17: 723-728. DOI: 10.1016/j.jcf.2018.02.008.

[68] Anuj SN, Whiley DM, Kidd TJ, Bell SC, Wainwright CE, Nissen MD, Sloots TP: Identification of *Pseudomonas aeruginosa* by a duplex real-time polymerase chain reaction assay targeting the *ecfX* and the *gyrB* genes. Diagn Microbiol Infect Dis. 2009; 63: 127-131. DOI: 10.1016/j.diagmicrobio.2008.09.018.

[69] Qin X, Emerson J, Stapp J, Stapp L, Abe P, Burns JL: Use of real-time PCR with multiple targets to identify *Pseudomonas aeruginosa* and other nonfermenting gram-negative bacilli from patients with cystic fibrosis. J Clin Microbiol. 2003; 41: 4312-4317. DOI: 10.1128/jcm.41.9.4312-4317.2003.

[70] Motoshima M, Yanagihara K, Fukushima K, Matsuda J, Sugahara K, Hirakata Y, Yamada Y, Kohno S, Kamihira S: Rapid and accurate detection of *Pseudomonas aeruginosa* by real-time polymerase chain reaction with melting curve analysis targeting *gyrB* gene. Diagn Microbiol Infect Dis. 2007; 58: 53-58. DOI: 10.1016/j.diagmicrobio.2006.11.007.

[71] Lavenir R, Jocktane D, Laurent F, Nazaret S, Cournoyer B: Improved reliability of *Pseudomonas aeruginosa* PCR detection by the use of the species-specific *ecfX* gene target. J Microbiol Methods. 2007; 70: 20-29. DOI: 10.1016/j.mimet.2007.03.008.

[72] DA Silva Filho LVF, Levi JE, Bento CNO, DA Silva Ramos SRT, Rozov

T: PCR identification of *Pseudomonas aeruginosa* and direct detection in clinical samples from cystic fibrosis patients. J Med Microbiol. 1999; 48: 357-361. DOI: 10.1099/00222615-48-4-357.

[73] Karpati F and Jonasson J: Polymerase chain reaction for the detection of *Pseudomonas aeruginosa, Stenotrophomonas maltophilia* and *Burkholderia cepacia* in sputum of patients with cystic fibrosis. Mol Cell Probes. 1996; 10: 397-403. DOI: 10.1006/mcpr.1996.0055.

[74] McCulloch E, Lucas C, Ramage G, Williams C: Improved early diagnosis of *Pseudomonas aeruginosa* by real-time PCR to prevent chronic colonisation in a paediatric cystic fibrosis population. J Cyst Fibros. 2011; 10: 21-24. DOI: 10.1016/j.jcf.2010.09.001.

[75] Le Gall F, Le Berre R, Rosec S, Hardy J, Gouriou S, Boisramé-Gastrin S, Vallet S, Rault G, Payan C, Héry-Arnaud G: Proposal of a quantitative PCR-based protocol for an optimal *Pseudomonas aeruginosa* detection in patients with cystic fibrosis. BMC Microbiol. 2013; 13: 143. DOI: 10.1186/1471-2180-13-143.

[76] Héry-Arnaud G, Nowak E, Caillon J, David V, Dirou A, Revert K, Munck MR, Frachon I, Haloun A, Horeau-Langlard D, Le Bihan J, Danner-Boucher I, Ramel S, Pelletier MP, Rosec S, Gouriou S, Poulhazan E, Payan C, Férec C, Rault G, Le Gal G, Le Berre R: Evaluation of quantitative PCR for early diagnosis of *Pseudomonas aeruginosa* infection in cystic fibrosis: a prospective cohort study. Clin Microbiol Infect. 2017; 23: 203-207. DOI: 10.1016/j.cmi.2016.11.016.

[77] Reyneke B, Ndlovu T, Khan S, Khan W: Comparison of EMA-, PMA- and DNase qPCR for the determination of microbial cell viability. Appl Microbiol Biotechnol. 2017; 101: 7371-7383. DOI: 10.1007/s00253-017-8471-6.

[78] Taylor SL, Leong LEX, Ivey KL, Wesselingh S, Grimwood K, Wainwright CE, Rogers GB; Australasian Cystic Fibrosis Bronchoalveolar Lavage (ACFBAL) study group: Total bacterial load, inflammation, and structural lung disease in paediatric cystic fibrosis. J Cyst Fibros. 2020; S1569-1993(20)30079-5. DOI: 10.1016/j.jcf.2020.03.008.

[79] Brauner A, Fridman O, Gefen O, Balaban NQ: Distinguishing between resistance, tolerance and persistence to antibiotic treatment. Nat Rev Microbiol. 2016; 14: 320-330. DOI: 10.1038/nrmicro.2016.34.

[80] Mahboubi MA, Carmody LA, Foster BK, Kalikin LM, VanDevanter DR, LiPuma JJ: Culture-Based and Culture-Independent Bacteriologic Analysis of Cystic Fibrosis Respiratory Specimens. J Clin Microbiol. 2016; 54: 613-619. DOI: 10.1128/JCM.02299-15.

[81] O'Toole GA: Cystic Fibrosis Airway Microbiome: Overturning the Old, Opening the Way for the New. J Bacteriol. 2018; 200: e00561-17. DOI: 10.1128/JB.00561-17.

[82] Marielle B, Sarrah G: Assessment of bacterial membrane fluidity by flow cytometry. J Microbiol Methods. 2017; 143: 50-57. DOI: 10.1016/j.mimet.2017.10.005.

[83] Rubio E, Zboromyrska Y, Bosch J, Fernandez-PittolMJ, FidalgoBI, FasanellaA, Mons A, Román A, Casals-Pascual C, Vila J: Evaluation of flow cytometry for the detection of bacteria in biological fluids. PLoS One. 2019; 14: e0220307. DOI: 10.1371/journal.pone.0220307.

[84] Lewenza S, Abboud J, Poon K, Kobryn M, Humplik I, Bell JR, Mardan L, Reckseidler-Zenteno S: *Pseudomonas aeruginosa* displays a dormancy phenotype during long-term survival in water. PLoS One. 2018; 13: e0198384. DOI: 10.1371/journal.pone.0198384.

[85] Dey R, Rieger AM, Stephens C, Ashbolt NJ: Interactions of *Pseudomonas*

aeruginosa with *Acanthamoeba polyphaga* Observed by Imaging Flow Cytometry. Cytometry A. 2019; 95: 555-564. DOI: 10.1002/cyto.a.23768.

[86] Narayanamurthy V, Sweetnam JM, Denner DR, Chen LW, Naureckas ET, Laxman B, White SR: The metabolic footprint of the airway bacterial community in cystic fibrosis. Microbiome. 2017; 5: 67. DOI: 10.1186/s40168-017-0289-z.

[87] Deschaght P, Schelstraete P, Lopes dos Santos Santiago G, Van Simaey L, Haerynck F, Van Daele S, De Wachter E, Malfroot A, Lebecque P, Knoop C, Casimir G, Boboli H, Pierart F, Desager K, Vaneechoutte M, De Baets F: Comparison of culture and qPCR for the detection of *Pseudomonas aeruginosa* in not chronically infected cystic fibrosis patients. BMC Microbiol. 2010; 10: 245. DOI: 10.1186/1471-2180-10-245.

[88] Boutin S, Weitnauer M, Hassel S, Graeber SY, Stahl M, Dittrich AS, Mall MA, Dalpke AH: Onetime quantitative PCR detection of *Pseudomonas aeruginosa* to discriminate intermittent from chronic infection in cystic fibrosis. J Cyst Fibros. 2018; 17: 348-355. DOI: 10.1016/j.jcf.2017.12.007.

[89] Amiri M, Mangiaterra G, Pasquaroli S, Vignaroli C, Cirilli N, Manso E, Biavasco F: Molecular diagnosis of *Pseudomonas aeruginosa* infection in culture-negative samples from cystic fibrosis patients. In: proceedings of 38th European Cystic Fibrosis Conference, J Cyst Fibros 2015, 14 S1: S73. DOI: 10.1016/S1569-1993(15)30239-3.

[90] Deschaght P, De Baere T, Van Simaey L, Van Daele S, De Baets F, De Vos D, Pirnay JP, Vaneechoutte M: Comparison of the sensitivity of culture, PCR and quantitative real-time PCR for the detection of *Pseudomonas aeruginosa* in sputum of cystic fibrosis patients. BMC Microbiol. 2009; 9: 244. DOI: 10.1186/1471-2180-9-244.

[91] Pasquaroli S, Citterio B, Mangiaterra G, Biavasco F, Vignaroli C: Influence of sublethal concentrations of vancomycin and quinupristin/dalfopristin on the persistence of viable but non-culturable *Staphylococcus aureus* growing in biofilms. J Antimicrob Chemother. 2018; 73: 3526-3529. DOI: 10.1093/jac/dky338.

[92] Mangiaterra G, Cedraro N, Vaiasicca S, Citterio B, Galeazzi R, Laudadio E, Mobbili G, Minnelli C, Bizzaro D, Biavasco F: Role of Tobramycin in the Induction and Maintenance of Viable but Non-Culturable *Pseudomonas aeruginosa* in an *In Vitro* Biofilm Model. Antibiotics (Basel). 2020; 9: 399. DOI: 10.3390/antibiotics9070399.

[93] Stanojevic S, Waters V, Mathew JL, Taylor L, Ratjen F: Effectiveness of inhaled tobramycin in eradicating *Pseudomonas aeruginosa* in children with cystic fibrosis. J. Cyst. Fibros. 2014; 13: 172-178. DOI: 10.1016/j.jcf.2013.09.002.

[94] Carvalhais V, Pérez-Cabezas B, Oliveira C, Vitorino R, Vilanova M, Cerca N: Tetracycline and rifampicin induced a viable nonculturable state in *Staphylococcus epidermidis* biofilms. Future Microbiol. 2018; 13: 27-36. DOI: 10.2217/fmb-2017-0107.

[95] Lee S and Bae S: Molecular viability testing of viable but non-culturable bacteria induced by antibiotic exposure. Microb. Biotechnol. 2018; 11: 1008-1016. DOI: 10.1111/1751-7915.13039.

[96] Soares A, Alexandre K, Etienne M: Tolerance and Persistence of *Pseudomonas aeruginosa* in Biofilms Exposed to Antibiotics: Molecular Mechanisms, Antibiotic Strategies and Therapeutic Perspectives. Front Microbiol. 2020; 11: 2057. DOI: 10.3389/fmicb.2020.02057.

Chapter 5

Pseudomonas aeruginosa Secreted Biomolecules and Their Diverse Functions in Biofilm Formation and Virulence

Theerthankar Das

Abstract

Pseudomonas aeruginosa is an opportunistic Gram-negative bacterium accountable for causing life-threatening infections in humans. According to the World Health Organization, *P. aeruginosa* classified as a critical pathogen. Specifically, *P. aeruginosa* in its colonized or biofilm state presents a major infection threat to immunocompromised (HIV) patients, Cystic fibrosis, burns, wounds and surgery associated infection. It is also a common pathogen responsible for causing hospital acquired/nosocomial infection and Urinary tract infections. *P. aeruginosa* biofilm is made up of bacterial self-synthesized biomolecules includes extracellular DNA, polysaccharides, proteins, RNA, siderophores and metabolites such as pyocyanin. This chapter will elaborate the manifold functions of *P. aeruginosa* secreted biomolecules in establishing and stabilizing biofilms, triggering virulence and pathogenicity in host, and resisting antibiotics and antibacterial agents.

Keywords: *Pseudomonas aeruginosa*, pyocyanin, extracellular DNA, biofilms, alginate, rhamnolipids, pyoverdine

1. Introduction

Pseudomonas aeruginosa is an opportunistic Gram-negative bacilli bacterium which holds a greater clinical significance in relation to its infection causing ability in humans [1]. *P. aeruginosa* is commonly found in environment (soil and water) and can be a source of contamination of drinking water and food spoilage [2, 3]. Prevalence of *P. aeruginosa* and its associated infection is commonly found in cystic fibrosis patient and chronic obstructive pulmonary disease (COPD) lungs, urinary tract, immunocompromised (HIV) patients, skin and soft-tissue, diabetic leg wounds, burns and surgical site infections [1, 4]. It is also a common pathogen responsible for causing healthcare associated (nosocomial) infection and microbial keratitis (eye infection due to contamination of contact lenses) [4]. World Health Organization (WHO) has listed *P. aeruginosa* as a most critical pathogen, due to the threat of causing blood stream infection (septicemia) and its antibiotic resistance ability [5]. *P. aeruginosa* or in general many other bacterial pathogens (e.g. *Acinetobacter baumannii*, *Escherichia coli*, *Klebsiella pneumoniae*, *Staphylococcus aureus*, *Streptococcus epidermidis*,

IntechOpen

Streptococcus pyogenes, *Proteus mirabilis*, etc.) has an inheritance ability to colonize and form biofilms on biotic (e.g. mucosa, tissue) or abiotic surface (e.g. medical implants, surgical instruments, hospital beds, wash basins, sinks, bath tub, etc.). Bacterial colonization on these surfaces directly leads to the contamination of surfaces, food and water and consequently precedes to infections in host. Bacterial biofilms are liable for approximately 80% of hospital and community-associated infections [6]. The most serious concern is antibiotic/antibacterial agents' resistance by the infecting bacteria that threatens the very core of modern medicine and impose a greater burden on global public health and economy.

2. Bacterial infection and antibiotic resistance are a global concern

National Institutes of Health (NIH), USA statistics data reports 550,000 death a year and about $94 billion total cost annually associated with biofilm infections in USA alone [7]. In Australia, billions of dollars expended annually dealing with antibiotic-resistant infections [8]. Statistics on antibiotic-resistance bacteria causing healthcare associated infections (HAI's) and death in European countries is alarming. Around, 8.9 million HAI recorded each year in combined hospitals and long-term care facilities and one in three bacteria associated with HAI's are antibiotic resistant [9]. In European population death associated with antibiotic resistance bacteria is estimated to be around 33,000 annually, this statistic is comparable to death associated with combined influenza, tuberculosis, HIV/AIDS [10]. Antibiotic resistance associated infections also triggers massive loss in productivity and healthcare incidentals estimated to be approximately 1.5 billion Euro's each year [11]. Bacterial resistance profile to different antibacterial agents is depends of several factors including geographical location of the strain (genetic mutations influenced by temperature, nutrient, oxygen), antibiotic prescribing practice to patients around the globe, poor hygiene and sanitation practice by common public and health care workers in hospitals. For instance, report released by Australian Commission on Safety and Quality in Health Care (Antimicrobial use and Resistance in Australia, AURA 2019) suggest more than 26.5 million antibiotic prescriptions were give out to patients [12]. The same report also highlighted that 23.5% of hospital prescribed antimicrobials in Australia are inappropriate and also community associated increase in antibiotic resistance bacteria (MRSA) are higher among old age people living in aged care facilities and in remote regions of the country [12]. Misuse or unethical use of antibiotics in agriculture, meat and poultry industry and fish farming, is a primary concern. Study published by Chinese Academy of Sciences reports use of 162,000 tons of antibiotics in the year 2013 alone in which more than half (52%) was used for animal husbandry and 48% by humans in addition, massive amount (50,000 tons) of antibiotics drained in the environment (water and soil) [13, 14]. USA also reported 10,000 tons of antibiotics annually used for livestock [15]. India, China, USA, Russia, Brazil, and South Africa are the world leaders in consuming antibiotics [16]. South China Morning Post (SCMP) Newspaper published an article in 2017 stating "Antibiotic overuse is a ticking time bomb for Asia" and health care workers act instantly to restrain misuse of antibiotics to stop public health calamity [17]. World Health Assembly, WHO, United Nations (UN), and countries respective government, local health organization and institutes are adopting a global action plan to crumb antibiotic resistance by educating common public, health care workers on effective sanitation, hygiene and infection prevention measures; and spreading awareness on responsible use of antibiotics in human and animal health, investing more funding in research and development in developing novel antibacterial agents, diagnostic tools, vaccines, improving hospital facilities especially in low-and-middle income countries [18].

3. *P. aeruginosa* Antibiotic resistance profile around the globe

P. aeruginosa inherent and adaptive antibiotic resistance character thus consequently making many existing antibiotics and anti-pseudomonal agents unusable against this bacterium and present a significant challenge for medical practitioners to treat infections. In this section, we exhibited few cases based on *P. aeruginosa* antibiotic resistance profile from different parts of the world by referring to previously published literature.

A comprehensive review by Wozniak et al., (2017), that covered the Australian data from the year 1990 till 2017, on antibiotic resistance Gram-negative bacteria [19]. Their study highlighted that *P. aeruginosa* isolates from different infection site showed resistance to many commonly used antibiotics. Among the *P. aeruginosa* isolates that were collected from surgical site between years 2002–2013, approximately 0.5%, 7.7% and 0.5% of the isolates showed resistance to fluoroquinolone, third generation cephalosporin and gentamicin respectively [20]. Survey on antibiotic resistance profile of *P. aeruginosa* isolates from patient's sputum between years 2007–2010 showed resistance to aminoglycosides (43%), beta-lactam (21%) and fluoroquinolone (30%) class of antibiotics [21]. Epidemiology studies on *P. aeruginosa* isolates from blood (years: 2001–2009) showed resistance to fluoroquinolone and meropenem about 12.7% and 14.3% respectively [22]. National Healthcare Safety Network (NHSN), USA survey on antimicrobial resistance patterns for the year 2009–2010, reported about 20% of pathogens (from 69,475 HAI's incidence) are antibiotic resistance in which 2% is carbapenem-resistant *P. aeruginosa* [23]. Microbial analysis on patients affected with Nosocomial and ventilator-associated pneumonia (VAP) in a period 2011–2012 in Georgia, USA reported *P. aeruginosa* as most prevalent Gram negative (40%) and highest prevalence of multi drug resistance [24]. Similar multi-drug resistance profile of *P. aeruginosa* was recorded in Asian countries. For example, antibiogram of total 2444 Pseudomonas species isolated from different clinical specimens (blood, pus, tracheal aspirate, urine and sputum from wards, intensive care units (ICUs) and follow up patients) of trauma patients from tertiary care hospitals in India over a period 2012–2016 revealed dominance of *P. aeruginosa* (95%) [25]. Among 69%, 68%, 67% 66%, 63% and 51% were levofloxacin, gentamicin, ciprofloxacin, ceftazidime, meropenem and tobramycin resistance, respectively [25]. Antibiotic profile of 121 *P aeruginosa* strains isolated from hospitals of Makkah and Jeddah, Saudi Arabia showed high resistance to antibiotics: meropenem (~30.6%), ticarcillin (22.3%), and imipenem (19%) [26]. A study reported that in mainland china hospitals prevalence of *P. aeruginosa* related ventilator-associated pneumonia (VAP) and hospital-acquired pneumonia were 19.4 and 17.8% respectively [27]. National Healthcare Safety Network (NHSN) USA, reports prevalence of *P. aeruginosa* is common among possible VAP [28]. These isolates exhibited high level of resistance to antibiotics: Gentamicin (up to 51.1%), cefoperazone (50%), and about 22.5% for amikacin [28]. *P. aeruginosa* resistance to ciprofloxacin has also risen a global concern, especially in Asian countries for example, Bangladesh reported 75.5% resistance to ciprofloxacin whereas, India, Iran, Turkey, and Saudi Arabia reported 49%, 58%, 48.9% and 50.9% respectively [29–33].

4. Role of *P. aeruginosa* secreted biomolecules in biofilm formation and virulence

Biofilm formation is the most preferred stage of many bacterial pathogens. Biofilm formation is a multi-step process to start with i) initial attachment of bacteria to the surface (adhesion) and to each other (aggregation), ii) growth regulations and microcolony formation and production of extracellular polymeric substances (EPS) and other exogenous molecules, iii) maturation of biofilms includes structural stability and iv) dispersal of bacterial cells from the mature biofilm into the environment and reestablishment at a new site [34].

Bacteria in its biofilm state are known to withstand antibacterial agents by many ten's and 100's-fold in comparison to its sessile/planktonic state [35]. Biofilm main composition includes up to 90% bacterially self-secreted biopolymers also known as extracellular polymeric substances (EPS) and other exogenous molecules and 10% bacterial cells [36]. These molecules in combined has been termed as house of bacteria and it shelter bacterial cells from numerous challenges includes antibiotics, antiseptics, detergents, shear mechanical stress, etc. [36]. Exogenous molecules synthesized by *P. aeruginosa* is primarily structured by a complex Quorum Sensing (QS) mechanism [37, 38]. In simple terms, QS is an intracellular communication phenomenon in which bacterial species able to detect and respond to its own cell population and ecological cues by regulating genes that facilitates them in survival and colonization in both biotic and abiotic environment. In *P. aeruginosa* QS is hierarchical and its driven through four known signaling system. At the top or first stage is driven by *las* system that activates the biosynthesis of autoinducing molecules N-(3-oxododecanoyl)-L-homoserine lactone (3-oxo-C12-HSL) and N-butanoyl-L-homoserine lactone (C4-HSL). Binding of LasR-HSL molecules triggers the transcription of second QS system: *rhlR*, *rhlI*, *lasI. LasR system further regulates the third and fourth:* 2-heptyl-3-hydroxy-4-quinolone *(HHQ) and* pseudomonas quinolone signal (PQS) [39]. These four QS circuits are interconnected and depends on each other regarding *P. aeruginosa* biosynthesis of various secreted and surface attached molecules. These includes extracellular biopolymers (Extracellular DNA, polysaccharides, proteins/enzymes), biosurfactant (rhamnolipids), metabolites (phenazine/pyocyanin), iron chelator (siderophore: pyoverdine, pyochelin), and bacterium cell surface anchored flagella and pili for swarming and twitching motilities [37–40]. These biomolecules and cell appendages independently or in coordination with each other plays dominant role in *P. aeruginosa* growth, fitness, biofilm formation, virulence, pathogenicity in host during infection, antibacterial resistance, and persistence. In this chapter we emphasize only on the diverse role of *P. aeruginosa* secreted extracellular biomolecules. **Figure 1** summarizes the diverse function of *P. aeruginosa* secreted extracellular biomolecules.

4.1 Extracellular DNA production, role in *P. aeruginosa* biofilm formation and stability

The role of extracellular (eDNA) in *P. aeruginosa* biofilm was first highlighted by Whitchurch et al. (2002) [41]. Their study revealed that eDNA is predominant in *P. aeruginosa* matrix component and its essential for *P. aeruginosa* biofilm formation [41]. Followed which numerous discoveries were done highlighting several roles of eDNA in *P. aeruginosa* and in other bacterial pathogens as well as in fungi [42–46]. Structural analysis study revealed that eDNA is similar to bacterial chromosomal DNA in its primary structure and it is not surprising because when chromosomal DNA release from bacterial cells (either via membrane vesicles or cell lysis) into

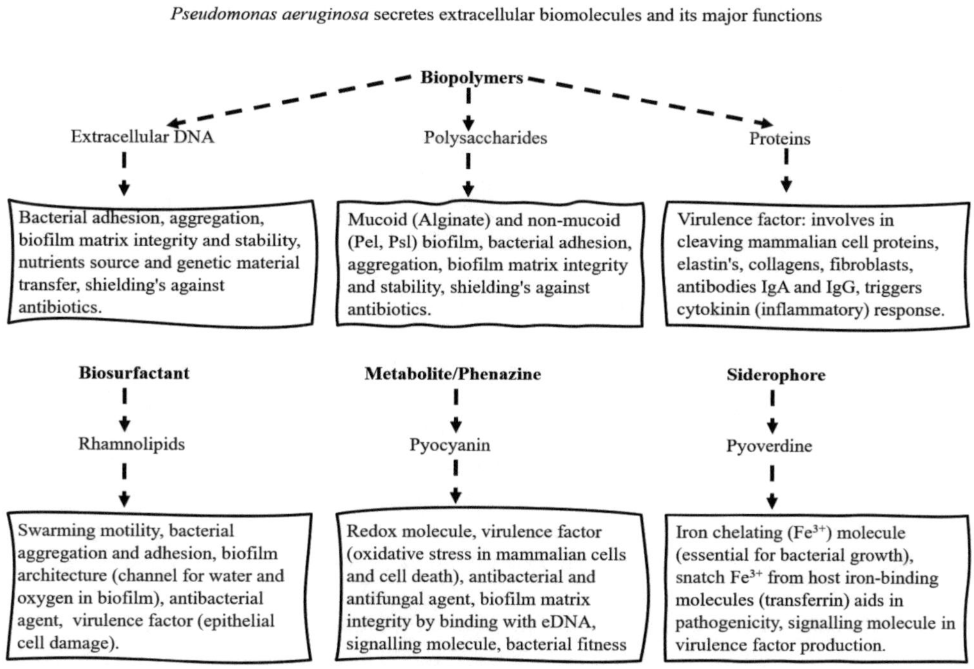

Figure 1.
Highlighting the major role of biomolecules secreted by P. aeruginosa. *These biomolecules are essential for establishment of biofilm, bacterial growth, fitness, and survival, induce virulence/pathogenicity and triggering immune response in host during infection, evading antibiotics, and other antibacterial agents.*

its immediate environment is termed as eDNA [47]. eDNA in *P. aeruginosa* cell population is released primarily through QS mechanism [48]. QS system (las and rhl -acyl homoserine lactone and pqs-*Pseudomonas* quinolone signaling), as well as flagella and type IV pili (*fliMpilA)* facilities prophage induction in *P. aeruginosa* cell population and consequently trigger cell lysis and eDNA release [48]. Virulence factor pyocyanin/phenazine biosynthesis also shown to trigger cell lysis (via oxidative stress mediated by hydrogen peroxide) and eDNA release in *P. aeruginosa* [49]. Outer membrane vesicles in *P. aeruginosa* cell also demonstrated to actively release eDNA [50].

Studies have confirmed that eDNA plays a key role in different stages of biofilm formation including initial bacterial to surface attachment (adhesion), bacteria-to-bacteria interaction (aggregation), colonization and biofilm formation by connecting cells to cells like nanowires [41–45]. Presence of eDNA on *P. aeruginosa* cell surface have shown to dictates physical surface properties of bacterial cell such as increase in cell surface hydrophobicity and consequently enables physico-chemical interactions forces such as Van der Waals interactions, Acid–Base interactions, hydrophobic interactions that aids in bacterial interactions and biofilm formation [51, 52]. eDNA have proven to induce electrostatic interactions with divalent cations like calcium (Ca^{2+}) and triggers bacterial aggregation [53]. eDNA has been established being an essential factor in structural integrity of *P. aeruginosa* biofilms and many studies have shown that cleaving of DNA using DNase I (enzyme that cleaves DNA through hydrolysis of phosphate di-ester bonds that links nucleotides in DNA) disrupts *P. aeruginosa* adhesion and biofilm formation [41, 44, 45, 54]. Other general roles of eDNA includes nutrient (e.g. good source of carbon, nitrogen, phosphorus) for starving bacteria and facilitate growth, horizontal gene transfer among bacteria cell (antibiotic resistance genes, virulence factor genes, etc), protects biofilms from shear stress by increasing biofilm viscosity. eDNA directly bindings to cationic antibiotics thus inhibits antimicrobial agents' interaction with bacteria within biofilm, removal of eDNA from biofilms have shown increase of bacterial susceptibility to antimicrobial agents [55]. In *P. aeruginosa* biofilm, eDNA release has shown to lower the pH of the local environment and subsequently these acidification initiates antibiotic resistance phenotype genes (PhoPQ and PmrAB) that fosters alteration of lipid A and the manufacture of spermidine on the *P. aeruginosa* outer membrane and consequently decrease entry/intake of aminoglycoside antibiotics [56].

4.2 Multitude task of polysaccharides secreted by *P. aeruginosa*

Many studies have concluded that polysaccharides as a chief component of many bacterial EPS/biofilm matrix. *P. aeruginosa* biosynthesis alginate, psl, and pel as their three predominant extracellular polysaccharides. Alginate producing isolates of *P. aeruginosa* have been acknowledged as a mucoid phenotype regulates through mutation in the alginate biosynthesis of *algA-algD* operon and *mucA* [57]. *AlgD* is the key gene that promotes alginate production followed by combined action of *mucA* and *algU* genes [57]. The physical characterizes of alginate positive *P. aeruginosa* colonizes are highly viscous and gelatinous structure on the edge of the cells [58]. This feature is due to its heavy molecular weight structure of alginate which mainly composed of O-acetylated D-mannuronic acid and its C5′ epimer L-guluronic acid [59]. Alginate productions make *P. aeruginosa* virulent strain and a foremost cause for respiratory infections and mortality in CF patients [60]. Alginate production enhances bacterial adhesion due to its sticky nature and its plays key role in shielding *P. aeruginosa* from host immune defense system by scavenging reactive oxygen species (ROS) and evading neutrophils and macrophages mediated phagocytosis [61, 62]. A study by McCaslin in rat alveolar macrophages, showed that alginate in combination with lipopolysaccharide produced by *P. aeruginosa* plays a synergy role in sparking airway inflammation by impeding alveolar function in removal of apoptotic cells and debris [63]. The anionic (negative charge) feature of alginate undergoes electrostatic interactions with cationic aminoglycosides and thus constrains their dissemination into biofilms [64]. Alginate also induce structural and conformational alteration and aggregation in the antimicrobial peptides by binding with it thereby, hinders its antimicrobial activity against pseudomonas [65].

In absence of alginate biosynthesize, *Psl* or *Pel* genes in *P. aeruginosa* isolates up-regulates and activates over production of psl and pel polysaccharide [58]. These polysaccharides by itself or in combination with each other exhibit non-mucoid bacterial colonies/biofilm and these colonies are termed as rugose small colony variant (RSCV) [58]. Psl biosynthesis in *P. aeruginosa* is induced through a QS (*las*) mediated set of *Psl* genes (*PslA-PslL*) and each or group of *Psl* genes and its corresponding protein/enzyme plays a unique role in synthesizing and integrating Psl polysaccharide [58]. For instance, PslB enzyme is responsible for sugar-nucleotide precursor production, whereas, PslA/PslE/PslJ/PslK/PslL and PslF, PslH, and PslI set of enzymes deals with polymerization of polysaccharide, and integration of the activated sugar subunits into the polysaccharide repeating structure [58]. Psl is a neutrally charged polysaccharide comprised of repeating sugar groups: D-mannose, L-rhamnose, and D-glucose [66, 67]. This polysaccharide plays a crucial role in bacterial cell-to-cell communication by enhancing intracellular c-di-GMP (secondary messenger molecule) and essential for initial *P. aeruginosa* attachment to a surface as tested on various clinical, environmental and common laboratory strains, biofilm biomass and antibiotic tolerance (tested on gentamicin) [68, 69].

Pel is a positively charged polysaccharide comprised of amino sugar groups and is biosynthesized is regulated via QS (*rhl* sytem) through activating *pel* operons (*pelA-pelG)* [70, 71]. Pel composed of acetylated 1–4 glycosidic linkages of N-acetylgalactosamine and N-acetylglucosamine [71]. PelA protein is responsible for the deacetylase of the sugar amino group, whereas PelD, PelE, PelF, and PelG enzymes, these set of enzymes accountable for Pel polymerization and passage across the *P. aeruginosa* cytoplasmic membrane [58, 71]. Study also speculated that pel is adapted version of LPS [71]. Pel polysaccharide biosynthesize is a strain dependent, and studies shown that in absence of psl polysaccharides pel genes up regulated to form primary structural framework in non-mucoid *P. aeruginosa* biofilms. This indicates that pel plays important role on later stage of biofilm and not during initial adhesion, aggregation, and colonization [58]. Pel being a cationic biopolymer binds to negatively charged eDNA in *P. aeruginosa* biofilm matrix via ionic bonding/electrostatic interactions henceforth, stabilize biofilm matrix frame [72].

4.3 *P. aeruginosa* exotoxins proteins role in pathogenicity

The biosynthesize and secretion of exogenous proteins/enzymes by *P. aeruginosa* is mediated by QS (*las-rhl*) system [73]. The common proteins virulence factor *P. aeruginosa* secrets includes elastase/LAS A and B, exotoxin A, U, S, T, Y phospholipase C, alkaline protease, type IV protease, phospholipase H and lipolytic enzymes [74]. The primary function of these proteins is to play as a virulence factor and induce bacterial pathogenicity in host. To induce pathogenicity, evade host immune defense and damage epithelial cells, *P. aeruginosa* secrets these proteins predominantly via type II and type III secretion system (out of five protein secretion system) [75, 76]. Type II system constituent of protein secretons that facilitate release of exotoxin A, elastase/LasA and LasB proteases, type IV protease, and phospholipase H, as well as lipolytic enzymes into the host cells. Whereas exotoxins U, S, T, and Y are released into host cells via type III secretion system (T3SS) [76]. T3SS forms needle like membrane structure that are anchored to the bacterial cell surface and facilitates delivery of bacterial protein virulence factors into the host epithelial cells [76].

Some actions of *P. aeruginosa* virulence proteins are discussed below. For example, *P. aeruginosa* toxin A protein have shown to impair protein elongation factor in mammalian cells thereby interferes with host essential protein synthesis [77]. The T3SS proteins (Exo U, S, T, Y) have diverse functions such as hinder DNA synthesis and modulates cell morphology in host, escaping host phagocytosis by impairing host cell actin cytoskeleton polymerization and endothelial barriers, phospholipase activity (cleaving host cell lipid layer and increase cell membrane permeability), modulates host inflammatory response and consequently extending bacteria and its virulence factors into host blood stream, different organs to cause bacteraemia and septicaemia and organ failure [78–80]. Metalloproteases are another group of enzymes such as elastase whose main function is to cleave human elastin and leukocyte elastase and neutrophil elastase and consequently alters host tissue elastic property and stimulate tissue damage. Elastases also proven to degrade human collagen II and IV, impair fibroblast growth and destroy wound healing proteins which are essential for mammalian cell and tissue development and wound repair [81–84]. Other crucial role of *P. aeruginosa* elastases includes cleaves host immunoglobulins (IgA and IgG) that aids bacterium to evade host immune response [85, 86]. Clinical studies in burn and wound patients infected with *P. aeruginosa*, showed protease biosynthesis by this bacterium trigger host cytokinin (interleukins IL6 and IL8) production and induce severe inflammation, septicaemia and elevates mortality level in patients [87–89].

4.4 Rhamnolipids *P. aeruginosa* biosurfactant

Rhamnolipids is a glycolipid biosurfactant produced by *P. aeruginosa* mediated through *rhl* QS system involving operons *rhlA*, *rhlB* for biosynthesis and *rhlI* and *rhlR* for regulation [90]. It is made up of sugar group (rhamnose) and a lipid/fatty acid group 3-(hydroxyalkanoyloxy) alkanoic acid and has a both hydrophilic and hydrophobic group like any typical biosurfactant [90, 91]. Rhamnolipids production helps *P. aeruginosa* in uptake and metabolism of hydrophobic molecules such as oils, hexadecane for nutritional source and growth [92]. Rhamnolipids (mono-rhamnolipids) also adhere to *P. aeruginosa* cell membrane (LPS) and plays key role in influencing *P. aeruginosa* cell surface physical property such as increasing cell surface hydrophobicity which aids in bacterial adhesion to substratum and bacterial cell-to-cell aggregation through hydrophobic interactions [93]. Rhamnolipids also lower the surface tension of *P. aeruginosa* cell surface thus aid them in swarming motility to travel across different location within the substratum [93]. It also proven to influence biofilm architecture by establishing and sustaining fluid channels in biofilms for water and oxygen transport [94]. *P. aeruginosa* employs rhamnolipids to their own advantage to eradicate competing bacteria. Binding of rhamnolipids into competing bacterial cell membrane consequently creates pores and increase cell permeability to induce cell lysis [95]. It is also a known virulence element, by binding to epithelial cell membrane it interrupts epithelial cell membrane integration, disrupts epithelial cell junctions, and triggers death in various mammalian cell types includes leukocytes, macrophages [96]. Rhamnolipids biosynthesis by *P. aeruginosa* in infected patients has been associated with escalation in pathogenicity in cystic fibrosis lung, ventilator-associated pneumonia patients [97].

4.5 Pyocyanin a unique virulence factor and its diverse function

P. aeruginosa biosynthesis and secretes a unique secondary metabolite called phenazines. Different types of phenazines are produced by *P. aeruginosa* however, pyocyanin is the most predominant one. Pyocyanin biosynthesis occurs at the later stage in *P. aeruginosa* population density or in biofilm, in laboratory culture it is generally expressed at the late exponential stage via regulation through QS (PQS) system [98]. Pyocyanin production is easily identified by its color, bluish -pure pyocyanin and green color when grown in laboratory in bacterial growth media (e.g. Tryptone Soy broth, Nutrient media, Luria broth, these media are all yellow in color and blue pyocyanin mix with yellow turns green). The two set genes of *phzA1-G1 and phzA2-G2* encrypts initial phenazine molecule (phenazine-1-carboxylic acid, PCA) followed by conversion of PCA to pyocyanin (N-methyl-1-hydroxyphenazine) encoded by genes *phzM* and *phzS* [98]. Pyocyanin production has been associated with the severity of infection and acknowledged as a hyper virulent strain [99]. Analysis of pyocyanin production on variety of clinical and environmental isolates indicates pyocyanin production is very common among all isolates however, the amount of pyocyanin production is depended upon strain phenotype and genotype variations. A study by Fothergill et al. (2007) on strains isolated from different clinical sites (CF, keratitis) and environmental (water) strains indicated that Liverpool epidermic strain (LES) from CF patients (attended Liverpool CF centre in England between years 1995 to 2004) exhibited significantly high pyocyanin production in comparison to keratitis and water isolates [99]. Pyocyanin plays diverse role in establishment of *P. aeruginosa* biofilm formation including inducing oxidative stress in competing bacteria and outcompete their growth (e.g. *S. aureus*) and fungi (e.g. *candida albicans*) [100, 101]. Pyocyanin promote cell signaling by activating transcription factor SoxR and stimulating various genes expression includes efflux pump genes *mexGHI-opmD*, and *PA2274* (monooxygenase, to control oxidative stress response in *P. aeruginosa*) [102]. By regulating target genes pyocyanin also maintain bacterial fitness, pyocyanin/phenazine deficient mutant (Δ*phz*) showed drastic change in its colony morphology (wrinkled colony), whereas pyocyanin over producing mutant strain (DKN370) remained smooth [103]. Pyocyanin induce oxidative stress and cell death (via H_2O_2 production) in *P. aeruginosa* population in late exponential phase and triggers eDNA production [49]. An interesting discovery by Das et al. 2012 and 2015 revealed that pyocyanin intercalates with DNA and influence *P. aeruginosa* cell surface hydrophobicity and subsequently promote biofilm formation [51, 104].

Pyocyanin has been in limelight in many decades due to its virulence property. In context to *P. aeruginosa* infection in human, pyocyanin production has been linked to increase in virulence and severity of infection [99]. Different studies reported different concentration of pyocyanin to be found in sputum of CF patients from 0.9 to 16.5 μg/ml and 27.3 μg/ml in bronchitis patients sputum and also significantly higher amount (5.3 μg/g) also found in burn wound exudates [105, 106]. In mammalian cells, it declines intracellular cAMP and ATP levels, provoke neutrophils apoptosis, and modulates host immune system [105–108]. Pyocyanin being a zwitter ion (positive and negative charge group and can penetrate into host cell membrane), and redox (electron donating and accepting property) molecule it oxidized cytosol (mammalian intracellular fluid), produces reactive oxygen species (ROS) by diffusing into host cells and undergoes redox reaction to accept electrons from NADPH and donates to molecular oxygen [109, 110]. ROS production leads to the decline in intracellular glutathione (a master antioxidant in mammalian cells essential for cell health and fitness) level which leads to bronchial epithelial cell death and tissue damage [109, 110]. It also impedes chlorine ion (Cl^-) secretion and transport in CF patients' lungs (bronchial epithelial cells) and consequently halt mucous clearance in human airways [111]. In burn wound patients infected with *P. aeruginosa*, pyocyanin production shown to provoke premature senescence and apprehend human fibroblast growth by levying oxidative stress [106, 112]. Mouse model study revealed that exposing pyocyanin to mouse lung airways triggers repress of transcription factors protein FoxA2 expression (essential for tissue development) and consequently leads to over production of host cells (cell hyperplasia) and mucous hypersecretion by [113].

4.6 Siderophore benefits *P. aeruginosa* growth and biofilm formation

Siderophore are small molecules and belongs to the class of "iron-chelating compounds". They are intrinsically secreted by microorganisms primarily for scavenging and uptake of Ferric ion, Fe^{3+} for their own benefits including nutrition, metabolism, growth, and virulence in mammals [114]. For example, Bacillus spp. *(subtilis and anthracis*) biosynthesis primary siderophore (bacillibactin), enterobactin, vibriobactin, yersinibactin, and pyoverdine by *E.coli*, *Vibrio cholerae*, *Yersinia pestis* and *P. aeruginosa* respectively. Pyoverdine is a fluorescent green color compound and its biosynthesis is encoded by the operons of *pvd*. Pyoverdine forages Fe^{3+} from host iron-binding molecules (transferrin) and binds strongly to it thus contribute to pathogenicity in host as shown in the immunocompromised mouse model [115, 116]. Pyoverdine also benefits from *P. aeruginosa* virulence factor protease action in degrading human iron-binding protein (ferritin), thus outcompetes host and scavenges iron [117]. Burn mouse model study have shown that pyoverdine contribute to severity in infection and mutants deficient in pyoverdine production showed significantly less virulence [116]. Infection model study in *Caenorhabditis elegans*, showed that pyoverdine penetrates host cells and undermines mitochondrial dynamics and triggers hypoxic response thus hinders ATP generation in host [118]. Other features of pyoverdine including communicating molecule to control biosynthesis of virulence proteins in *P. aeruginosa* including exotoxin A and protease [119]. Iron is essential to sustain bacterial growth thus pyoverdine aids in survival of *P. aeruginosa* in infection site, triggers biofilm formation where, pyoverdine deficient mutant strains forms fragile biofilm [120]. *P. aeruginosa* also produces another siderophore molecule called pyochelin, however pyochelin has lower affinity for Fe^{3+} than pyoverdine. However, this pyochelin-iron complex in coordination with pyocyanin undergoes oxidative-reductive reaction and contribute to oxidative damage (via hydroxyl radical formation) and inflammation in host [121, 122]. In CF patients pyochelin found to be involved in inflammation and tissue damage [123].

5. Conclusion

P. aeruginosa ability to easily colonize in host, biofilm formation, synthesis and secretion of virulence factors and causing pathogenicity, evading host immune defense system, and antimicrobial resistance made it a critical pathogen and needs an immediate attention. Secretion of extracellular molecules by *P. aeruginosa* plays a principal role in fitness of bacterial population, establishment of biofilms, infections, and pathogenicity in host. To reduce and eradicate *P. aeruginosa* associated infections development of novel antibiotics or antimicrobial agents, QS inhibiting molecules, virulence factor neutralizing agents, biofilm disrupting enzymes or/and combination treatment strategy with existing antimicrobial agents are of top priority. Further to prevent antimicrobial resistance in bacteria, necessary steps need to be taken by government organization, hospitals/clinics, health care workers and scientist from research institutes to educate children and students in schools, colleges and universities, people from rural places and developing countries about proper hygiene and use and misuse of antibiotics. Also, proper management of antibiotics uses in the agriculture and meat industry need to be implemented. This small steps at every level will help in minimize the spread of antimicrobial resistance in bacteria and will help to cut down catastrophe in both health and economic division and promotes better treatment outcome against infectious diseases.

Author details

Theerthankar Das
Department of Infectious Diseases and Immunology, School of Medical Sciences, The University of Sydney, Australia

*Address all correspondence to: das.ashishkumar@sydney.edu.au

References

[1] Centers for diseases control and prevention. Available from: https://www.cdc.gov/hai/organisms/pseudomonas.html

[2] Kouchesfahani MM, Alomo hammadi M, Nodehi RN, Aslani H, Rezaie S, Asadian S. *Pseudomonas aeruginosa* and *Heterotrophic* Bacteria Count in Bottled Waters in Iran. Iran J Public Health.2015;44(11):1514-1519.

[3] Arslan S, Eyi A, Ozdemir F. Spoilage potentials and antimicrobial resistance of Pseudomonas spp. isolated from cheeses. J. Dairy Sci. 2011;94 :5851-5856. doi: 10.3168/jds.2011-4676.

[4] [Internet]. Available from: https://www.msdmanuals.com/professional/infectious-diseases/gram-negative-bacilli/pseudomonas-and-related-infections

[5] [Internet]. Available from: https://www.who.int/news/item/27-02-2017-who-publishes-list-of-bacteria-for-which-new-antibiotics-are-urgently-needed.

[6] Davies D. Understanding biofilm resistance to antibacterial agents. Nat Rev Drug Discov. 2003;2(2):114-122.

[7] Wolcott RD, Rhoads DD, Bennett ME, Wolcott BM, Gogokhia L, Costerton JW, Dowd SE. Chronic wounds and the medical biofilm paradigm. J Wound Care. 2010;19(2):45-46, 48-50, 52-43.

[8] [Internet]. Available from: Respiratory infectious disease burden in Australia. (2007), 1 edn. The Australian Lung Foundation.

[9] [Internet]. Available from: https://antibiotic.ecdc.europa.eu/en/publications-data/healthcare-associated-infections-threat-patient-safety-europe.

[10] [Internet]. Available from: https://antibiotic.ecdc.europa.eu/en/publications-data/antibiotic-resistance-increasing-threat-human-health.

[11] The bacterial challenge: Time to react. (2009). European Centre for Disease Prevention and Control/European Medicines Agency.

[12] [Internet]. Available from: https://www.safetyandquality.gov.au/our-work/antimicrobial-resistance/antimicrobial-use-and-resistance-australia-surveillance-system-aura/aura-2019

[13] [Internet]. Available from: http://english.cas.cn/newsroom/archive/news_archive/nu2015/201507/t20150715_150362.shtml

[14] [Internet]. Available from: https://www.theguardian.com/environment/2018/jun/19/can-china-kick-its-animal-antibiotic-habit].

[15] [Internet]. Available from: https://www.theguardian.com/environment/2018/jun/19/can-china-kick-its-animal-antibiotic-habit.

[16] Van Boeckel TP, Gandra S, Ashok A, Caudron Q, Grenfell BT, Levin SA, Laxminarayan R. Global antibiotic consumption 2000 to 2010: An analysis of national pharmaceutical sales data. Lancet Infect Dis. 2014;14(8):742-750.

[17] [Internet]. Available from: https://www.scmp.com/comment/insight-opinion/article/2119765/antibiotic-overuse-ticking-time-bomb-asia.

[18] [Internet]. Available from: World Health Organization (2015) Global Action Plan on AntimicrobialResistance. http://apps.who.int/iris/bitstream/10665/193736/1/9789241509763_eng.pdf.

[19] Wozniak TM, Paterson D, Halton K. Review of the epidemiological data regarding antimicrobial resistance in Gram-negative bacteria in Australia. Infect Dis & Health. 2017;22(4):210-218. Doi.org/10.1016/j.idh.2017.07.003

[20] Worth LJ, Bull AL, Spelman T, Brett J, Richards MJ. Diminishing Surgical Site Infections in Australia: Time Trends in Infection Rates, Pathogens and Antimicrobial Resistance Using a Comprehensive Victorian Surveillance Program, 2002-2013. Infect Cont & Hosp Epidem. 2015;36(4):409-416. DOI: https://doi.org/10.1017/ice.2014.70.

[21] Smith DJ, Ramsay KA, Yerkovich ST, Reid DW, Wainwright CE, Grimwood K, Bell SC, Kidd TJ. Pseudomonas aeruginosa antibiotic resistance in Australian cystic fibrosis centres. Respirology. 2016 Feb;21(2):329-37. Doi: 10.1111/resp.12714.

[22] Anug AK, Skinner MJ, Lee FJ, Cheng AC. Changing epidemiology of bloodstream infection pathogens over time in adult non-specialty patients at an Australian tertiary hospital. Commun Dis Intell Q Rep. 2012;36(4):E333-41.

[23] Sievert DM, Ricks P, Edwards JR, Schneider A, Patel J, Srinivasan A, Kallen A, Brandi Limbago B, Fridkin S, Antimicrobial-resistant pathogens associated with healthcare-associated infections: summary of data reported to the National Healthcare Safety Network at the Centers for Disease Control and Prevention, 2009-2010. Infect Control Hosp Epidemiol. 2013. 34(1):1-14. Doi: 10.1086/668770.

[24] Behnia M, Logan SC, Linda Fallen L, Catalano P. Nosocomial and ventilator-associated pneumonia in a community hospital intensive care unit: a retrospective review and analysis. BMC Res Notes. 2014;7:232. Doi: 10.1186/1756-0500-7-232.

[25] Kumari M, Khurana S, Bhardwaj N, Malhotra R, Mathur P. Pathogen burden & associated antibiogram of Pseudomonas spp. in a tertiary care hospital of India. Indian J Med Res. 2019 Feb;149(2):295-298. Doi: 10.4103/ijmr.IJMR_14_18.

[26] Khan MA, Faiz A. Antimicrobial resistance patterns of *Pseudomonas aeruginosa* in tertiary care hospitals of Makkah and Jeddah.Ann Saudi Med. 2016;36(1):23-8. Doi: 10.5144/0256-4947.2016.23.

[27] Ding C, Yang Z, Wang J, Liu X, Cao Y, Pan Y, Han L, Zhan S. Prevalence of *Pseudomonas aeruginosa* and antimicrobial-resistant *Pseudomonas aeruginosa* in patients with pneumonia in mainland China: a systematic review and meta-analysis. Internat J of Infect Dis. 49 (2016) 119-128120. DOI: https://doi.org/10.1016/j.ijid.2016.06.014

[28] [Intranet] Available from: https://www.cdc.gov/nhsn/faqs/faq-vae.html#q14

[29] Rashid A, Chowdhury A, Rahman SHZ, Begum SA, Muazzam N. Infections by *Pseudomonas aeruginosa* and antibiotic resistance pattern of the isolates from Dhaka Medical College hospital. Bangladesh J Med Microbiol. 2007;1:48-51.

[30] Rajat Rakesh M, Ninama Govind L, Kalpesh M, Rosy P, Kanu P, Vegad MM. Antibiotic resistance pattern in *Pseudomonas aeruginosa* species isolated at a tertiary care hospital, Ahmadabad. Nat J Med Res. 2012;2:156-159.

[31] Golshani Z, Ahadi AM, Sharifzadeh A. Antimicrobial susceptibility pattern of *Pseudomonas aeruginosa* isolated from patients

referring to hospitals. Arch Hyg Sci. 2012;1:48-53.

[32] Savas L, Duran N, Savas N, Onlen Y, Ocak S. The prevalence and resistance patterns of *Pseudomonas aeruginosa* in intensive care units in a university hospital. Turk J Med Sci. 2005;35:317-322.

[33] Asghar AH, Faidah HS. Frequency and antimicrobial susceptibility of gram-negative bacteria isolated from 2 hospitals in Makkah, Saudi Arabia. Saudi Med J. 2009;30:1017-23.

[34] Maunders E, Welch M. Matrix exopolysaccharides; The sticky side of biofilm formation. 2017: FEMS Microbiol Letters 364(13), fnx120 doi: 10.1093/femsle/fnx120

[35] Carol Potera ANTIBIOTIC RESISTANCE: Biofilm Dispersing Agent Rejuvenates Older Antibiotics. Environ Health Perspect. 2010 Jul; 118(7): A288.

[36] Flemming HC, Wingender J. The biofilm matrix. Nature Reviews Microbiology. 2010;8: 623-633.

[37] Parsek MR, Greenberg EP. Socio microbiology: The connections between quorum sensing and biofilms. Trends in microbiology. 2005;13(1):27-33.

[38] Jimenez PN, Koch G, Thompson JA, Xavier KB, Cool RH, Quax WJ. The multiple signalling systems regulating virulence in *Pseudomonas aeruginosa*. Microbiol Mol Biol Rev. 2012;76(1):46-65.

[39] Lee J, Zhang L. The hierarchy quorum sensing network in *Pseudomonas aeruginosa*. Protein Cell. 2015; 6(1):26-41. Doi: 10.1007/s13238-014-0100-x.

[40] Davies DG, Parsek MR, Pearson JP, Iglewski BH, Costerton JW, Greenberg EP. The involvement of cell-to-cell signals in the development of a bacterial biofilm. Science. 1998;280(5361):295-298.

[41] Whitchurch CB, Tolker-Nielsen T, Ragas PC, Mattick JS. Extracellular DNA required for bacterial biofilm formation. Science. 2002 22;295(5559):1487. Doi: 10.1126/science.295.5559.1487.

[42] Izano EA, Amarante MA, Kher WB, Kaplan JB. Differential roles of poly-*N*-acetylglucosamine surface polysaccharide and extracellular DNA in *Staphylococcus aureus* and *Staphylococcus epidermidis* biofilms. Appl Environ Microbiol. 2008;74(2):470-476.

[43] Harmsen M, Lappann M, Knochel S, Molin S. Role of extracellular DNA during biofilm formation by *Listeria monocytogenes*. Appl Environ Microbiol. 2010;76(7):2271-2279.

[44] Das T, Sharma PK, Busscher HJ, Van Der Mei HC, Krom BP. Role of extracellular DNA in initial bacterial adhesion and surface aggregation. Appl Environ Microbiol. 76 (10), 3405-3408

[45] Das T, Krom BP, Van der Mei HC, Busscher HJ, Sharma PK. DNA-mediated bacterial aggregation is dictated by acid–base interactions. Soft Matter 7 (6), 2927-2935.

[46] Shopova I, Bruns S, Thywissen A, Kniemeyer O, Brakhage AA, Hillmann F. Extrinsic extracellular DNA leads to biofilm formation and colocalizes with matrix polysaccharides in the human pathogenic fungus *Aspergillus fumigatus*. Front Microbiol. 2013. https://doi.org/10.3389/fmicb.2013.00141

[47] Steinberger RE, Holden PA. Extracellular DNA in single- and multiple-species unsaturated biofilms. Appl Environ Microbiol. 2005. 71:5404-10.

[48] Allesen-Holm M, Barken KB, Yang L, Klausen M, Webb JS,

Kjelleberg S, Molin S, Givskov M, Tolker-Nielsen T. A characterization of DNA release in *Pseudomonas aeruginosa* cultures and biofilms. Mol Microbiol. 2006;59(4):1114-1128.

[49] Das T, Manefield M. Pyocyanin promotes extracellular DNA release in *Pseudomonas aeruginosa*. PLoS One. 2012;7(10):e46718.

[50] Kadurugamuwa JL, Beveridge TJ. Virulence factors are released from *Pseudomonas aeruginosa* in association with membrane vesicles during normal growth and exposure to gentamicin: A novel mechanism of enzyme secretion. J Bacteriol. 1995;177(14):3998-4008.

[51] Das T, Kutty SK, Kumar N, Manefield M. Pyocyanin facilitates extracellular DNA binding to *Pseudomonas aeruginosa* influencing cell surface properties and aggregation. PLoS One. 2013;8(3):e58299.

[52] Kimyon Ö, Das T, Ibugo AI, Kutty SK, Ho KK, Tebben J, Kumar N, Manefield M. *Serratia* secondary metabolite prodigiosin inhibits *Pseudomonas aeruginosa* biofilm development by producing reactive oxygen species that damage biological molecules. Front in Microbiol. 2016;7(972).

[53] Das T, Sehar S, Koop L , Wong YK, Ahmed S , Siddiqui KS, Manefield M. Influence of calcium in extracellular DNA mediated bacterial aggregation and biofilm formation. PLoS One. 2014 Mar 20;9(3):e91935. Doi: 10.1371/journal.pone.0091935. eCollection 2014.

[54] Swartjes JJTM, Das T, Sharifi S, Subbiahdoss G, Sharma PK, Krom BP, Busscher HJ, Van der Mei HC. A functional DNase I coating to prevent adhesion of bacteria and the formation of biofilm. Adv Function Material.2012;23:2843-2849. DOI:10.1002/adfm.201202927.

[55] Purdy Drew KR, Sanders LK, Culumber ZW, Zribi O, Wong GC. Cationic amphiphiles increase activity of aminoglycoside antibiotic tobramycin in the presence of airway polyelectrolytes. J Am Chem Soc. 2009;131(2):486-493.

[56] Wilton M, Charron-Mazenod L, Moore R, Lewenza S. Extracellular DNA acidifies biofilms and Iiduces aminoglycoside resistance in *Pseudomonas aeruginosa*. Antimicrob Agents Chemother. 2015;60(1):544-553.

[57] Pulcrano G, Vita Lula D, Raia V, Rossano F, Catania MR. Different mutations in mucA gene of *Pseudomonas aeruginosa* mucoid strains in cystic fibrosis patients and their effect on algU gene expression. New Microbiol. 2012 Jul;35(3):295-305.

[58] Franklin MJ, David E, Nivens DE, Joel T. Weadge JT, P. Lynne Howell PL. Biosynthesis of the *Pseudomonas aeruginosa* extracellular polysaccharides, alginate, Pel, and Psl. Front. Microbiol., 22 August 2011 | https://doi.org/10.3389/fmicb.2011.00167.

[59] Linker, A., and Jones, R. S. (1964). A polysaccharide resembling alginic acid from a *Pseudomonas* micro-organism. *Nature* 204, 187-188.

[60] Govan JR, Deretic V. Microbial pathogenesis in cystic fibrosis: Mucoid *Pseudomonas aeruginosa* and *Burkholderia cepacia*. Microbiol Rev. 1996;60(3):539-574.

[61] Pier GB (1998) *Pseudomonas aeruginosa*: A key problem in cystic fibrosis. 64 (6):339-347.

[62] McCaslin CA, Daniela N Petrusca DN, Poirier C, Karina A Serban KA, Gregory G Anderson GG, Petrache I. Impact of alginate-producing *Pseudomonas aeruginosa* on alveolar macrophage apoptotic cell clearance.

J Cyst Fibros. 2015 Jan;14(1):70-77. doi: 10.1016/j.jcf.2014.06.009.

[63] Nichols WW, Dorrington SM, Slack MP, Walmsley HL. Inhibition of tobramycin diffusion by binding to alginate. Antimicrob Agents Chemother. 1988;32(4):518-523.

[64] Chan C, Burrows LL, Deber CM. Helix induction in antimicrobial peptides by alginate in biofilms. J Biol Chem. 2004;279(37):38749-38754.

[65] Friedman L, Kolter R. Two genetic loci produce distinct carbohydrate-rich structural components of the *Pseudomonas aeruginosa* biofilm matrix. J Bacteriol. 2004;186(14):4457-4465.

[66] Byrd MS, Sadovskaya I, Vinogradov E, Lu H, Sprinkle AB, et al. Genetic and Biochemical Analyses of the *Pseudomonas aeruginosa* Psl Exopolysaccharide Reveal Overlapping Roles for Polysaccharide Synthesis Enzymes in Psl and LPS Production. Mol Microbiol. 2009 Aug; 73(4): 622-638. Doi: 10.1111/j.1365-2958.2009.06795.x

[67] Irie Y, Borlee BR, O'Connor JR, Hill PJ, Harwood CS, et al. Self-produced exopolysaccharide is a signal that stimulates biofilm formation in *Pseudomonas aeruginosa.* PNAS 2012;109 (50) 20632-20636; https://doi.org/10.1073/pnas.1217993109

[68] Irie Y, Roberts AEL, Kragh KN, Gordon VD, Hutchison J, et al. The *Pseudomonas aeruginosa* PSL Polysaccharide Is a Social but Noncheatable Trait in Biofilms. DOI: 10.1128/mBio.00374-17.

[69] Ryder C, Byrd M, Wozniak DJ. Role of polysaccharides in *Pseudomonas aeruginosa* biofilm development. Curr Opin Microbiol. 2007;10(6):644-648.

[70] Coulon C, Vinogradov E, Filloux A, Sadovskaya I (2010) Chemical analysis of cellular and extracellular carbohydrates of a biofilm-forming strain *Pseudomonas aeruginosa* PA14. PLoS One 5(12):e14220.

[71] Colvin KM, Irie Y, Tart CS, Urbano R, Whitney JC. The Pel and Psl polysaccharides provide *Pseudomonas aeruginosa* structural redundancy within the biofilm matrix. Environ Microbiol. 2012 Aug; 14(8): 10.1111/j.1462-2920.2011.02657.x.

[72] Jennings LK, Storek KM, Ledvina HE, Coulon C, Marmont LS, et al. Pel is a cationic exopolysaccharide that cross-links extracellular DNA in the *Pseudomonas aeruginosa* biofilm matrix. PNAS September 8, 2015 112 (36) 11353-11358. Doi: https://doi.org/10.1073/pnas.1503058112

[73] Nouwens AS, Beatson SA, Whitchurch CB, Walsh BJ, Herbert HP, et al. Proteome analysis of extracellular proteins regulated by the las and rhl quorum sensing systems in *Pseudomonas aeruginosa* PAO1. Microbiology (Reading). 2003;149(Pt5):1311-1322.doi: 10.1099/mic.0.25967-0.

[74] Cianciotto NP. Type II secretion: a protein secretion system for all seasons. Trends Microbiol. 2005;13:581-8.

[75] Jyot J, Balloy V, Jouvion G, Verma A, Touqui L. Type II Secretion System of *Pseudomonas aeruginosa*: In Vivo Evidence of a Significant Role in Death Due to Lung Infection. J Infect Dis. 2011 May 15; 203(10): 1369-1377. doi: 10.1093/infdis/jir045.

[76] Engel J, Balachandran P. Role of *Pseudomonas aeruginosa* type III effectors in disease. Curr Opin Microbiol. 2009;12:61-6.

[77] Iglewski BH, Kabat D. NAD-dependent inhibition of protein synthesis by *Pseudomonas aeruginosa* toxin. Proc Natl Acad Sci U S A. 1975;72(6):2284-2288.

[78] Sato H, Frank DW. ExoU is a potent intracellular phospholipase. Mol Microbiol. 2004;53: 1279-1290.

[79] Galle M, Carpentier I, Beyaert R. Structure, and function of the Type III secretion system of *Pseudomonas aeruginosa*. Curr Protein Pept Sci. 2012;13(8):831-842. doi: 10.2174/138920312804871210.

[80] Hauser AR. The type III secretion system of *Pseudomonas aeruginosa*: infection by injection. Nat Rev Microbiol. 2009;7(9):654-665. doi: 10.1038/nrmicro2199.

[81] Peters JE, Park SJ, Darzins A, Freck LC, Saulnier JM, Wallach JM, Galloway DR. Further studies on *Pseudomonas aeruginosa* LasA: Analysis of specificity. Molecular Microbiology. 1992;6(9):1155-1162.

[82] Schmidtchen A, Holst E, Tapper H, Bjorck L. Elastase-producing *Pseudomonas aeruginosa* degrade plasma proteins and extracellular products of human skin and fibroblasts, and inhibit fibroblast growth. Microb Pathog. 2003;34(1):47-55.

[83] Kessler E, Safrin M. Synthesis, processing, and transport of *Pseudomonas aeruginosa* elastase. J Bacteriol. 1988;170(11):5241-5247.

[84] Heck LW, Morihara K, McRae WB, Miller EJ. Specific cleavage of human type III and IV collagens by *Pseudomonas aeruginosa* elastase. Infect Immun. 1986;51(1):115-118.

[85] Heck LW, Alarcon PG, Kulhavy RM, Morihara K, Russell MW, Mestecky JF. Degradation of IgA proteins by *Pseudomonas aeruginosa* elastase. J Immunol. 1990;144(6): 2253-2257.

[86] Bainbridge T, Fick RB, Jr. Functional importance of cystic fibrosis immunoglobulin G fragments generated by *Pseudomonas aeruginosa* elastase. J Lab Clin Med. 1989;114(6):728-733.

[87] Murray PR. Manual of clinical microbiology. 1999. ASM Press, Washington.

[88] Holder IA, Haidaris CG. Experimental studies of the pathogenesis of infections due to *Pseudomonas aeruginosa*: Extracellular protease and elastase as in vivo virulence factors. Canadian Journal of Microbiology. 1979;25(5):593-599.

[89] Voynow JA, Fischer BM, Zheng S. Proteases and cystic fibrosis. Int J Biochem Cell Biol. 2008;40(6-7):1238-1245.

[90] Desai JD, Banat I M. Microbial production of surfactants and their commercial potential. Microbiol Mol Biol Rev. 1997 Mar; 61(1): 47-64.

[91] Rikalovic MG, Structure-Function Relationships of Rhamnolipid and Exopolysacharide Biosurfactants of *Pseudomonas aeruginosa* as Therapeutic Targets in Cystic Fibrosis Lung Infections. IntechOpen, 2017.

[92] Noordman WH, Janssen DB. Rhamnolipid Stimulates Uptake of Hydrophobic Compounds by *Pseudomonas aeruginosa*. Appl Environ Microbiol. 2002 Sep; 68(9): 4502-4508. Doi: 10.1128/AEM.68.9.4502-4508.2002.

[93] Zhong H, Zeng GM, Yuan XZ, Fu HY, Huang GH, Ren FY. Adsorption of dirhamnolipid on four microorganisms and the effect on cell surface hydrophobicity. Appl Microbiol Biotechnol. 2007;77 (2): 447-55. doi:10.1007/s00253-007-1154-y.

[94] Davey ME, Caiazza NC, O'Toole GA. "Rhamnolipid surfactant production affects biofilm architecture in *Pseudomonas aeruginosa* PAO1. J Bacteriol. 2003;185(3):1027-36. doi:10.1128/jb.185.3.1027-1036.2003.

[95] Sotirova AV, Spasova DI, Galabova DN, Karpenko E, Shulga A. Rhamnolipid-biosurfactant permeabilizing effects on gram-positive and gram-negative bacterial strains. Curr Microbiol. 2008;56 (6): 639-44. doi:10.1007/s00284-008-9139-3.

[96] Zulianello L, Canard C, Köhler T, Caille D, Lacroix JS, Meda P. Rhamnolipids are virulence factors that promote early infiltration of primary human airway epithelia by *Pseudomonas aeruginosa*. Infect. Immun. 2006;74 (6): 3134-47. doi:10.1128/IAI.01772-05. PMC 1479292.

[97] Köhler T, Guanella R, Carlet J, van Delden C. "Quorum sensing-dependent virulence during *Pseudomonas aeruginosa* colonisation and pneumonia in mechanically ventilated patients". Thorax. 2010;65 (8): 703-10. doi:10.1136/thx.2009.133082.

[98] Mavrodi DV, Bonsall RF, Delaney SM, Soule MJ, Phillips G, Thomashow LS. Functional analysis of genes for biosynthesis of pyocyanin and phenazine-1-carboxamide from *Pseudomonas aeruginosa* PAO1. J Bacteriol. 2001;183(21):6454-6465.

[99] Fothergill JL, Panagea S, Hart CA, Walshaw MJ, Tyrone L Pitt TL, Winstanley C. Widespread pyocyanin over-production among isolates of a cystic fibrosis epidemic strain. BMC Microbiol. 2007; 7: 45. Doi: 10.1186/1471-2180-7-45.

[100] Sweden EG. Study the effect of antibiotics on pyocyanin production from *Pseudomonas aeruginosa* and pyocyanin as antibiotic against different pathogenic bacteria. Journal of University Anbar Pure Science. 2010; 4:15-18

[101] Kerr JR, Taylor GW, Rutman A, Hoiby N, Cole PJ, Wilson R. Pseudomonas aeruginosa pyocyanin and 1-hydroxyphenazine inhibit fungal growth. Journal of Clinical Pathology. 1999; 52:385-387.

[102] Dietrich LEP, Price-Whelan A, Petersen A, Whiteley M, Newman DK. The phenazine pyocyanin is a terminal signalling factor in the quorum sensing network of *Pseudomonas aeruginosa*. 2006;61(5):1308-1321.

[103] Dietrich LEP, Teal TK, Price-Whelan A, Newman DK Redox-Active Antibiotics Control Gene Expression and Community Behavior in Divergent Bacteria. Science. 2008; 321(5893): 1203-1206.

[104] DasT,KuttySK,TavallaieR,IbugoAI, Panchompoo J, Sehar S, Aldous L, Yeung AWS, Thomas SR, Kumar N, Gooding JJ, Manefield M. Phenazine virulence factor binding to extracellular DNA is important for *Pseudomonas aeruginosa* biofilm formation. 2015;5(8398).

[105] Wilson R, Sykes DA, Watson D, Rutman A, Taylor GW, Cole PJ. Measurement of *Pseudomonas aeruginosa* phenazine pigments in sputum and assessment of their contribution to sputum sol toxicity for respiratory epithelium. Infection and Immunity. 1988;56(9):2515-2517.

[106] Muller M, Li Z, Maitz PK. *Pseudomonas* pyocyanin inhibits wound repair by inducing premature cellular senescence: role for p38 mitogen-activated protein kinase. Burns. 2009;35(4):500-508.

[107] Munro NC, Barker A, Rutman A, Taylor G, Watson D, McDonald-Gibson WJ, Towart R, Taylor WA, Wilson R, Cole PJ. Effect of pyocyanin and 1-hydroxyphenazine on in vivo tracheal mucus velocity. J Appl Physiol (1985). 1989;67(1):316-323.

[108] Denning GM, Wollenweber LA, Railsback MA, Cox CD, Stoll LL, Britigan BE. *Pseudomonas* pyocyanin

increases interleukin-8 expression by human airway epithelial cells. Infect Immun. 1998;66(12):5777-5784.

[109] O'Malley YQ, Reszka KJ, Spitz DR, Denning GM, Britigan BE. *Pseudomonas aeruginosa* pyocyanin directly oxidizes glutathione and decreases its levels in airway epithelial cells. Am J Physiol Lung Cell Mol Physiol. 2004;287(1):L94-103.

[110] Rada B, Jendrysik MA, Pang L, Hayes CP, Yoo DG, Park JJ, Moskowitz SM, Malech HL, Leto TL. Pyocyanin-enhanced neutrophil extracellular trap formation requires the NADPH oxidase. PLoS One. 2013;8(1):e54205.

[111] Schwarzer C, Fischer H, Kim EJ, Barber KJ, Mills AD, et al. Oxidative Stress By Pyocyanin Impairs CFTR Cl^- Transport In Human Bronchial Epithelial Cells. Free Radic Biol Med. 2008; 45(12): 1653-1662. Doi: 10.1016/j.freeradbiomed.2008.09.011.

[112] Gonzalez MR, Fleuchot B, Lauciello L, Jafari P, Applegate LA, Raffoul W, Que YA, Perron K. Effect of human burn wound exudate on *Pseudomonas aeruginosa* virulence. mSphere. 2016;1(2).

[113] Hao Y, et al. *Pseudomonas aeruginosa* pyocyanin causes airway goblet cell hyperplasia and metaplasia and mucus hypersecretion by inactivating the transcriptional factor FoxA2. Cellular microbiology. 2012;14:401-415].

[114] Challis GL (April 2005). "A widely distributed bacterial pathway for siderophore biosynthesis independent of nonribosomal peptide synthetases". ChemBioChem. 6 (4): 601-11. doi:10.1002/cbic.200400283. PMID 15719346.

[115] Sriyosachati S, Cox CD. Siderophore-mediated iron acquisition from transferrin by *Pseudomonas aeruginosa*. Infect Immun. 1986;52(3):885-891.

[116] Meyer JM, Neely A, Stintzi A, Georges C, Holder IA. Pyoverdin is essential for virulence of *Pseudomonas aeruginosa*. Infection and Immunity. 1996;64(2):518-523.

[117] Doring G, Pfestorf M, Botzenhart K, Abdallah MA. Impact of proteases on iron uptake of *Pseudomonas aeruginosa* pyoverdin from transferrin and lactoferrin. Infect Immun. 1988;56(1):291-293.

[118] Kirienko, Natalia V.; Kirienko, Daniel R.; Larkins-Ford, Jonah; Wählby, Carolina; Ruvkun, Gary; Ausubel, Frederick M. (2013-04-17). "*Pseudomonas aeruginosa* disrupts *Caenorhabditis elegans* iron homeostasis, causing a hypoxic response and death". Cell Host & Microbe. 13 (4): 406-416. doi:10.1016/j.chom.2013.03.00.

[119] Lamont IL, Beare PA, Ochsner U, Vasil AI, Vasil ML. Siderophore-mediated signaling regulates virulence factor production in *Pseudomonas aeruginosa*. Proc Natl Acad Sci U S A. 2002;99(10):7072-7077.

[120] Lamont IL, Konings AF, Reid DW. Iron acquisition by *Pseudomonas aeruginosa* in the lungs of patients with cystic fibrosis. Biometals. 2009;22(1):53-60.

[121] Coffman TJ, Cox, CD, Edeker BL., and Britigan BE. Possible role of bacterial siderophores in inflammation. Iron bound to the Pseudomonas siderophore pyochelin can function as a hydroxyl radical catalyst. *J. Clin. Invest*. 1990;86, 1030-1037. doi: 10.1172/JCI114805.

[122] Britigan BE, Rasmussen G., Cox CD. Augmentation of oxidant injury to human pulmonary epithelial

cells by the *Pseudomonas aeruginosa* siderophore pyochelin. *Infect. Immun*. 1997;65, 1071-1076.

[123] Lyczak JB, Cannon CL,Pier GB. Lung infections associated with cystic fibrosis. *Clin Microbiol Rev*. 2002;15, 194-222. doi: 10.1128/CMR.15.2.194-222.2002.

Chapter 6

Chemotherapy and Mechanisms of Action of Antimicrobial Agent

Rahman Laibi Chelab

Abstract

Pseudomonas aeruginosa is a widespread opportunistic pathogen that causes bloodstream, urinary tract, burn wounds infections and is one of the largest pathogens that infect cystic fibrosis patients' airways and can be life-threatening for *P. aeruginosa* infections. In addition, *P. aeruginosa* remains one of the most significant and difficult nosocomial pathogens to handle. Increasingly, multi-drug resistance (MDR) strains are identified and the option of therapy is often very limited in these cases, particularly when searching for antimicrobial combinations to treat serious infections. The fact that no new antimicrobial agents are active against the MDR strains of *P. aeruginosa* is an additional matter of concern. In recent decades, bacterial drug resistance has increased, but the rate of discovery of new antibiotics has decreased steadily. The fight for new, powerful antibacterial agents has therefore become a top priority. This chapter illustrates and explores the current state of several innovative therapeutic methods that can be further discussed in clinical practice in the treatment of *P. aeruginosa* infections.

Keywords: *P. aeruginosa*, drug resistance, alternative therapies, vaccine, phage therapy

1. Introduction

We are currently facing an international crisis with many troublesome aspects: new antibiotics are no longer being detected, resistance mechanisms are developing in almost all clinical isolates of bacteria, and the effective treatment of infections is hampered by recurrent infections caused by persistent bacteria. Antibiotic failure is one of the most worrying health issues worldwide [1]. Although resistance acquisition is a natural phenomenon, it is accelerated by antibiotic misuse, inadequate inspection and poorly regulated management of antibiotics have resulted in the appearance and spread of multidrug-resistant (MDR) bacteria abroad in clinical medicine and in the livestock industry [2, 3].

Empirical antibiotic treatment requires monotherapy and combination therapy for suspected cases of *P. aeruginosa* and reduces mortality in patients with serious *P. aeruginosa* infections [4, 5]. However, because of the ability of this bacterium to avoid many of the currently available antibiotics, treatment of *P. aeruginosa* infections has become a great challenge [6]. Recently, the World Health Organization (WHO) has identified carbapenem-resistant *P. aeruginosa* as one of three bacterial species with an urgent need for new antibiotics to be developed to treat infections [7]. In addition, inappropriate treatment use of antibiotics accelerates the production of multidrug-resistant strains of *P. aeruginosa*, resulting in the ineffectiveness of empirical antibiotic therapy against this microorganism [8]. Resistance to a range of antibiotics, including aminoglycosides, quinolones and β-lactams, is demonstrated by *P. aeruginosa* [9]. Generally, the main mechanisms of *P. aeruginosa* used to fight antibiotic attack can be divided into intrinsic, acquired and adaptive resistance. Low external membrane permeability, the expression of efflux pumps that remove antibiotics from the cell, and the development of antibiotic inactivating enzymes are part of the intrinsic resistance of *P. aeruginosa*. Either horizontal transfer of resistance genes or mutational modifications will achieve the acquired resistance of *P. aeruginosa* [10]. *P. aeruginosa* 's adaptive resistance requires the development of biofilm in the lungs of infected patients, where the biofilm functions as a diffusion barrier to inhibit the access of antibiotics to bacterial cells [11].

The effectiveness and safety of murepavadin in the treatment of infections of the lower respiratory tract caused by *P. aeruginosa* (suspected or confirmed) in patients with ventilation-associated pneumonia or CF-unrelated bronchiectasis (Clinical Trials. gov identifiers NCT02096315 and NCT02096328) have been tested in two clinical trials. However, by July 17, 2019, the trials were stopped because in research participants who had obtained murepavadin, an unusually high level of renal failure had been found. This decision would not impact the production of an aerosolized formulation of murepavadin for topical use [12]. Murepavadin is a particular weapon against *P. aeruginosa*, which separates it from the broad pipeline of antimicrobial natural and synthetic peptides acting against multiple taxa, *P. aeruginosa* included. Recently, several novel peptides with broad antimicrobial activity have been identified, such as antimicrobial peptide DGL13K, Mel4 and melamine (Melimine and Mel4 are chimeric cationic peptides with broad-spectrum antimicrobial activity), Cecropin B, Lysine-based peptidomimetics (LBP-2), Truncated pseudin-2 analogs (Pse-T2), antimicrobial peptide, termed 6 K-F17 (sequence: KKKKKK-AAFAAWAAFAA-NH_2), Melittin-derived peptides (MDP1, MDP2) [13–20]. In addition, multidrug-tolerant persistent cells can form in the biofilm that are capable of surviving antibiotic attack; in cystic fibrosis (CF) patients, these cells are responsible for prolonged and recurrent infections [21]. For patients whose infections are resistant to traditional antibiotics, the development of new antibiotics or alternative therapeutic methods for treating *P. aeruginosa* infections is urgently needed. In recent years, new antibiotics with novel modes of action have been investigated, as have new routes of administration and resistance to bacterial enzyme alteration. Compared to traditional antibiotics, some of these newer antibiotics demonstrate excellent in vitro antibacterial activity against *P. aeruginosa* as well as lower minimum inhibitory concentration (MIC) [22, 23]. Moreover, several novel non-antibiotic therapeutic approaches that are highly successful in destroying antibiotic resistant *P. aeruginosa* strains have been documented in recent studies [24]. These approaches include: antimicrobial peptides, phage therapy, inhibition of quorum sensing, iron chelation, the use of nanoparticles, probiotic and vaccine strategy. In order to combat *P. aeruginosa* infections, these therapeutic approaches may be used either as an alternative to or in conjunction with traditional antibiotic therapies.

2. Mechanisms of action of antimicrobial agents

There are six basic mechanisms of antimicrobial agents presented below:

1. Inhibition of microbial cell wall synthesis.

2. Inhibition of microbial cell membrane function.

3. Inhibition of microbial protein synthesis.

4. Inhibition of microbial DNA synthesis.

5. Inhibition of microbial RNA synthesis.

6. Inhibition of microbial metabolic pathways.

2.1 Inhibition of microbial cell wall synthesis

An integral microbial structure responsible for the shape of the cell is the cell wall. In addition, because of the high cytoplasmic osmotic pressure, the cell wall prevents cell lysis and facilitates the anchoring of membrane components and extracellular proteins, such as adhesins [25]. Bacterial cell wall synthesis has perhaps become the target field most commonly exploited for antimicrobial production on the basis of the number of antimicrobial drugs in clinical usage. Due to the absence of equivalents in human biology, the components of the cell wall synthesis machinery are attractive antimicrobial targets, thus providing intrinsic objective selectivity. The cytoplasmic synthesis of building blocks composed of N-acetyl muramic acid (M) linked to N-acetyl glucosamine (G) with an attached pentapeptide (P) side chain (referred to as MGP subunits) comprises the sequential late steps in cell wall synthesis. The linkage of the MGP subunit to the lipid II molecule enables subsequent translocation to the outside or periplasmic space of the cell through the cytoplasmic membrane. By catalyzing glycosidic linkages between the M and G components of the MGP subunits, transglycosylase enzymes then assemble the MGP subunits into a linear backbone. An immature peptidoglycan structure is constituted by linearly connected MGP subunits. Transpeptidase enzymes then work to cross-link pentaglycine bridges to the peptide side chains, in the process, the terminal 2 D-alanines of the peptide side chain are cleaved, creating the mature, lattice-like peptidoglycan that provides the form and osmotic stability of the bacterium [26]. β-lactam antibiotics, such as penicillins and cephalosporins, are the most widely used antimicrobials that prevent cell wall biosynthesis [27]. These β-lactam antibiotics interact directly with bacterial transpeptidases and inhibit them effectively. As transpeptidase inhibitors, β-lactams thus obstruct the transition from immature to mature peptidoglycan, so these enzymes are also referred to as penicillin-binding proteins (PBPs). Due to the stereochemical similarity of the β-lactam moiety with the D-alanine-D-alanine substrate, they are capable of doing this. Transpeptidases form a lethal covalent penicilloyl enzyme complex in the presence of the drug, which helps to inhibit the usual transpeptidation reaction. This results in peptidoglycan that is weakly cross-linked, which makes the developing bacteria extremely susceptible to cell lysis and death [28].

2.2 Inhibition of microbial cell membrane function

Lipids, proteins and lipoproteins are essentially made of biological membranes. The cytoplasmic membrane for water, ions, nutrients and transport systems serves as a diffusion barrier. Most health workers now assume that membranes are a lipid matrix with uniformly distributed globular proteins to penetrate through the bilayer of the lipid. A number of antimicrobial agents may cause disorganization of the membrane. These agents can be categorized into cationic, anionic, and neutral agents. Polymyxin B and colistemethate (polymyxin E) are the best-known compounds [29]. For several antimicrobial agents, the cytoplasmic membrane forms an important barrier.

The mode of action of certain antimicrobial agents may be due to the ability of such medicines to increase membrane permeability, making it easier for them and other compounds to penetrate. Antibacterial cationic agents, increased permeability of the outer membrane to the lysozyme and hydrophobic compounds has been identified, such as polymyxin B. The initial function of these antimicrobial agents is to interrupt the structure of the outer membrane, allowing the cell to join itself and other compounds and inhibit unique metabolic processes [30]. There are several cell-damaging properties of Polymixin B: (i) the surface charge, lipid composition and membrane structure are disturbed; (ii) the K+ gradient on the cytoplasmic membrane is dissipated; and (iii) the cytoplasmic membrane is depolarized. One of the key factors regulating bacterial exposure to polymixin B is the permeability of the external membrane to lipophilic compounds. Since polymixin B is bulkier than its displacement of inorganic divalent cations, in the presence of polymixin B, the packing order of lipopolysaccharides (LPS) is changed. This results in increased permeability of a variety of molecules to the outer membrane and also promotes polymixin B uptake ("self-promoted" uptake) [31].

2.3 Inhibition of microbial protein synthesis

Microbial protein synthesis inhibition a range of groups of antimicrobial agents work by inhibiting the synthesis of bacterial proteins (ribosome function). That include aminoglycosides, macrolides, tetracyclines, ketolides, lincosamides, streptogramins, chloramphenicol and oxazolidinones [26, 32]. The synthesis of microbial proteins is led by ribosomes in conjunction with cytoplasmic factors which, during the initiation phase, elongation phase and termination phases, bind transiently to particles. Microbial ribosomes contain 70S particles consisting of two 50S and 30S subunits, which join at the initiation stage of the synthesis of proteins and split at the termination stage. In bacterial protein synthesis, antimicrobial agents block various steps by interfering with the work of either the cytoplasmic factors or the ribosomes. Inhibitors which bind to the ribosomal subunit of 30S primarily interfere with initiation, although some interfere with the pairing of the AA- tRNA anticodon with the mRNA codon, elongation is thus impaired. The steps involved in the elongation process interact with inhibitors that bind to the 50S ribosomal subunit or to elongation factors that are transiently connected to ribosomes at certain stages of the cycle.

Through binding to particular ribosomal subunits [33], aminoglycosides function. By inducing the development of aberrant, non-functional complexes as well as causing misreading, aminoglycoside-type drugs may combine with other binding sites on 30S ribosomes and destroy bacteria. Spectinomycin is an antimicrobial agent that is closely linked to the aminoglycosides of aminocylitol. It binds and is bacteriostatic but not bactericidal to a particular protein in the ribosome. Tetracyclines are other agents which bind to 30S ribosomes. These agents tend to inhibit aminoacyl tRNA binding to the A site of the bacterial ribosome. Tetracycline binding is temporary, so it's bacteriostatic for these agents. Nevertheless, a wide range of bacteria, *chlamydias* and *mycoplasmas* are inhibited and highly helpful agents [29]. There are three major groups of medicines that inhibit the ribosomal subunit of 50S. A bacteriostatic agent that inhibits both gram-positive and gram-negative bacteria is chloramphenicol. By binding to a peptidyltransferase enzyme on the 50S ribosome, it prevents peptide bond formation. Macrolides are large compounds of the lactone ring that bind to 50S ribosomes and tend to impair the reaction or translocation of peptidyltransferase, or both. Erythromycin, which inhibits gram-positive species and a few gram-negative species, such as *haemophilus, mycoplasma, chlamydia* and *legionella*, is the most significant macrolide. Against many of these pathogens, new molecules including azithromycin and clarithromycin have greater activity than erythromycin. There is a similar activity site for lincinoids, the most important of which is clindamycin. Generally, macrolides and lincinoids are bacteriostatic and only inhibit the development of new peptide chains [29].

2.4 Inhibition of microbial DNA synthesis

The modulation of chromosomal supercoiling by topoisomerase-catalyzed strand breakage and rejoining reactions is needed for DNA synthesis, mRNA transcription and cell division [34]. Depending on whether they catalyze reactions involving transient breakage of one (type I) or both (type II) strands of DNA, DNA topoisomerase enzymes are classified into two groups, I and II [35]. The topological state of DNA inside cells is regulated by topoisomerases and is important for the vital processes of protein translation and cell replication. The enzyme that negatively super-coils DNA in the presence of ATP is DNA gyrase, a type II DNA topoisomerase [36]. Moreover, in the absence of ATP, this enzyme plays a role in the catenation and decatenation reaction of a double-stranded DNA circle, resolves knots in DNA, and also relaxes supercoiled DNA negatively. As a result, for almost all cellular procedures involving duplex DNA, including replication, recombination and transcription, the enzyme is vital. It is unique to the prokaryotic kingdom and is essential to the organism's survival. Thus, for antibacterial drugs, DNA gyrase remains an ideal and attractive target. The most effective DNA gyrase-targeted antimicrobial agents are quinolones. Nalidixic acid, a naphthyridone inadvertently discovered as a by-product during chloroquine synthesis, was the source of the compounds [37].

Quinolones are unique DNA-gyrase inhibitors. DNA gyrase reactions such as supercoiling and relaxation involving DNA breakage and reunion are inhibited by quinolones, specifically interfering with the DNA gyrase breakage-reunion reaction by interacting with subunit A (GyrA) [38]. Relatively poor antimicrobial activity is found in first-generation quinolones, nalidixic acid and oxolinic acid. However, the synthesis and improvement over many generations of fluoroquinolones, such as norfloxacin and ciprofloxacin (second generation), levofloxacin (third generation), and moxifloxacin and gemifloxacin (fourth generation), has resulted in a variety of potent antimicrobial agents [38]. Most bacterial pathogens possess an additional essential topoisomerase, topoisomerase I (Topo I), in addition to the type II topoisomerases. Topo I is architecturally and mechanistically distinct from gyrase and topoisomerase IV, and is an attractive candidate for new antibacterial chemotypes to be discovered as such [36].

2.5 Inhibition of microbial RNA synthesis

Rifamycins inhibit DNA-dependent transcription by binding the DNA-bound and effectively transcribing RNA polymerase with a high affinity to the β-subunit (coded by rpoB). In the channel formed by the RNA polymerase-DNA complex, from which the newly synthesized RNA strand emerges, the β- subunit is located. Rifamycins clearly require that RNA synthesis has not progressed beyond two ribonucleotides being added; This is due to the drug molecule 's capacity to sterically inhibit the initialization of nascent RNA strands. It should be noted that rifamycins are not believed to work by blocking the RNA synthesis elongation stage, although a recently discovered class of RNA polymerase inhibitors (based on the CBR703 compound) could inhibit elongation by modifying the enzyme allosterically [34].

2.6 Inhibition of microbial metabolic pathways

By competitively blocking the biosynthesis of tetrahydrofolate, which acts as a carrier of one-carbon fragments and is required for the ultimate synthesis of DNA, RNA and bacterial cell wall proteins, trimethoprim and sulfonamides interfere with folic acid metabolism in the microbial cell. Bacteria and protozoan parasites typically lack a transport mechanism in order to extract preformed folic acid from their host, unlike mammals [29]. Most of these species, while some are capable of using exogenous thymidine, must synthesize folic acid, circumventing the need for metabolism of folic acid. The conversion of pteridine and p-aminobenzoic acid (PABA) to dihydrofolic acid by the pteridine synthetase enzyme is competitively inhibited by sulfonamides. Sulfonamides have a greater affinity for pteridine synthetase than for PABA. Trimethoprim has a huge affinity (10,000 to 100,000 times greater than that of the mammalian enzyme) for bacterial dihydrofolate reductase; it inhibits tetrahydrofolate synthesis when bound to this enzyme [29].

3. Mechanisms of resistance to antimicrobial agents

The ability of an organism to overcome the action of an antimicrobial agent to which it was previously susceptible is a general definition of antimicrobial resistance [39]. With the growing production of MDR strains (i.e. resistance to at least three antibiotics), nosocomial infection caused by antibiotic resistant *P. aeruginosa* has emerged as a major concern in clinical care settings [40]. Because of its outer membrane with low permeability (1/100 of the permeability of the outer membrane of *E. coli*), *P. aeruginosa* exhibits intrinsic resistance to various antimicrobial agents (β-lactam and penem group of antibiotics) [41, 42]. While several other processes are also responsible for their intrinsic resistance, including the efflux system that expels antibiotics from the cell's bacteria and the production of inactivating enzyme antibiotics. This bacterium, however is a highly diverse pathogen capable of adapting to the conditions around it. When subjected to selective pressure from antibiotics, the mediated reaction encourages bacterial survival and improves resistance to antibiotics [43–45].

The development of antibiotic resistance during host colonization of patients with CF has been confirmed, with *P. aeruginosa* strains developing and gaining resistance during antimicrobial therapy [46]. Studies have shown a clear link between increased applications of ciprofloxacin, with a growing incidence of strains resistant to ciprofloxacin [47]. Therefore the excessive use of antimicrobial agents is another factor associated with the rise in MDR-*Psedomonas aeruginosa*. This acquired resistance may be attributable to the effects of the mutational event or the acquisition by horizontal gene transfer of the resistance gene and may occur during the mutational event of antibiotic therapy, leading to over-expression of endogenous β-lactamases or efflux pump, specific porin expression [48].

3.1 Resistance to β-lactam

Inhibition of the synthesis of the bacterial peptidoglycan cell wall requires β-lactam antibiotics [39]. Penicillin, cephalosporin, carbapenem and monobactam are included in this class. These classes include piperacillin and ticarcillin (penicillin), ceftazidime (cephalosporin 3rd generation), cefepime (cephalosporin 4th generation), aztreonam (monobactam), imipenem, meropenem and doripenem (carbapenems) are most powerful β-lactam widely used to treat *P. aeruginosa* is β-lactam [49]. These enzymes break the amide bond of the β-lactam ring through resistance to the β-lactam mediated by the action of β-lactamases, rendering the antimicrobial ineffective. The expression of endogenous β-lactamases or the expression of acquired β-lactamases may be due to this inactivation of the drug. To date, hundreds of β-lactamases have been recognized and are distinguished by their substrate specificity. There are four major groups of beta-lactamases known in *P. aeruginosa* on the basis of Amber's molecular classification system: A-D [50]. Through the serine-residue catalytic activity, classes A, C and D inactivate the β-lactams, while class B or metallo- β-lactamases (MBLs) require zinc for their action [51].

3.2 AmpC β-lactamase (Cephelosporinase)

In particular, the development of endogenous β-lactamase, such as chromosomal cephalosporinase (AmpC β-lactamase). A variety of β-lactams, such as benzyl penicillin, narrow spectrum cephalosporin and imipenem, can be induced in *P. aeruginosa*. Naturally, *P. aeruginosa* is susceptible to carboxypenicillins, ceftazidime and aztreonam, but it can develop resistance through a mutation in the gene that contributes to AmpC β-lactamase hyper-production [52, 53]. The enzyme is usually produced in small amounts ('low-level' expression), resistance to aminopenicillins and to most early cephalosporins is determined. *P. aeruginosa* produces an inducible chromosome-coded AmpC β-lactamase (cephalosporinase) belonging to the Ambler-based molecular class C and the first functional group according to Bush et al. [54, 55].

However, production of chromosomal cephalosporin in *P. aeruginosa*, in the presence of inducing β-lactams (especially imipenem), can increase from 100 to 1000 times [56]. β-lactamase inhibitors used in clinical practice, such as clavulanic acid, sulbactam and tazobactam, do not inhibit AmpC cephalosporinase function. β-lactamase of AmpC is encoded by the gene ampC [57, 58]. Several genes, including ampR, ampG, and ampD, are involved in ampC gene induction. AmpR encodes a positive transcriptional regulator and this regulator is required for the induction of β-lactamase. AmpG, a transmembrane protein that functions as a permease for 1,6-anhydromurapeptides, which are known to be the signal molecules involved in the induction of ampC, is the second gene involved. The third gene, ampD, encodes a cytosolic amidase of N-acetyl-anhydromuramyl-L-alanine that hydrolyses 1,6-anhydromurapeptides, which functions as an ampC expression repressor. The 4th chromosome, ampE, encodes the protein of the cytoplasmic membrane that serves as the molecule of the sensory transducer necessary for induction. Except for avibactam, the activity of this AmpC β-lactamase is not inhibited by commercially available β-lactam.

3.3 Class A carbenicillin hydrolysing β-lactamases

Four β-lactamases (PSE- of *Pseudomonas* specific enzyme) carbenicillin hydrolyzing enzymes were identified in *P. aeruginosa*: PSE-1 (CARB-2), PSE-4 (CARB-1), CARB-3 and CARB-4 [59]. These enzymes belong to the group of β-lactamases of molecular class A and include carboxypenicillins, ureidopenicillins and cefsulodine in their substrate profile. These enzymes belong to functional group 2c and molecular class A [60]. Commercially available β-lactam inhibitors, such as clavulanic acid, tazobactam, and sulbactam, can inhibit the activity of this β-lactamase [61].

3.4 Resistance to aminoglycoside

Aminoglycosides are a microbial protein synthesis inhibitor which act by binding to the ribosomal subunit of the bacterial 30S and interfering with the initiation of protein synthesis. Resistance to aminoglycosides in *Pseudomonas* is mediated by transferable aminoglycoside modifying enzymes (AMEs), low permeability of the outer membrane, active efflux and, in rare cases, target modification [62–64].

3.5 Aminoglycoside-modifying enzymes

AMEs inactivate the aminoglycoside by adding the antibiotic molecule to a phosphate, adenyl or acetyl radical, and thus modified antibiotics minimize the binding affinity of the bacterial cell (30S ribosomal subunit) to its target [65, 66]. Aminoglycoside phosphoryl transferases (APHs), aminoglycoside adenylyl transferases (also known as nucleotidyltransferases) (AADs or ANTs) and aminoglycoside acetyltransferases (AACs) are three types of AMEs involved in aminoglycoside alteration. The following AMEs are most commonly expressed by *P. aeruginosa:* AAC(69)-II (resistant to gentamicin, tobramycin and netilmicin), AAC(3)-I (resistant to gentamicin), AAC(3)-II (resistant to gentamicin, tobramycin and netilmicin), (69)-I (resistant to tobramycin, netilmicin and amicacin) and ANT(29)-I (resistant to tobramicin and gentamicin) [67].

3.6 Low outer membrane permeability

Membrane impermeability or reduced permeability is a mechanism known to provide resistance to many antibiotic forms, including aminoglycosides, β-lactams and quinolones [68]. For instance, this resistance mechanism is often encountered in cystic fibrosis isolates that are continually under antibiotic attack. Several mechanisms, such as lipopolysaccharide (LPS) modifications, alteration of membranous proteins involved in substratum absorption, and inactivation of enzymatic complexes involved in the energetic membrane necessary for transport system activity, may cause membrane impermeability [69].

3.7 Active efflux pumps

The combination of low membrane permeability and active efflux pumps is partially due to the natural resistance of *P. aeruginosa* to many groups of antibiotics. *P. aeruginosa's* efflux systems involved in antibiotic resistance belong to the family of resistance-nodulation-division (RND) [70]. In order to confer resistance to several antibiotics, four major efflux systems have been described: MexAB-OprM, MexCD-OprJ, MexEF-OprN and MexXY-OprM. These systems consist of three proteins: (1) the efflux pump protein found in the cytoplasmic membrane (MexB, MexD, MexF and MexY), (2) the pore-acting outer membrane protein (OprM, OprJ and OprN) and (3) A protein in the periplasmic space that bridges the cytoplasmic and outer membrane proteins (MexA, MexC, MexE and MexX). In both natural and acquired resistance, MexAB-OprM and MexXY-OprM are active, whereas only the other two mechanisms are observed in cumulative resistance.

Acquired resistance is observed following mutations in the regulatory systems that can be caused by antibiotic pressure and that can confer resistance to all groups of antibiotics upon over-expression of these efflux systems. Polymyxins, except [69]. Resistance to multiple groups of antibiotics that are substrates of these efflux systems can be caused by exposure to a single antibiotic. Quinolones are substrates of all efflux systems and are an important trigger factor that can generate cross-resistance to efflux systems of several major classes of antibiotics, including β-lactams and aminoglycosides, for pseudomonal therapy [71]. It is understood that efflux systems confer a moderate degree of resistance, but they typically act simultaneously with other mechanisms of resistance, thus taking part in the high-level resistance that can be observed in *P. aeruginosa.*

3.8 Target modification

Due to the low affinity of the drug to the bacterial ribosome, bacteria may be resistant to aminoglycosides. This can be achieved by 16S rRNA methylation by target modification. Various 16S rRNA methylases have been identified for *P. aeruginosa*: RmtA, first reported in clinical isolates of *P. aeruginosa* resistant to aminoglycosides and conferred resistance to all parenterally administered aminoglycosides, including amicacin, tobramycin, isepamicin, kanamycin, arbecacin and gentamicin, secondary 16S rRNA methylases including RmtB, ArmA and RmtD [72].

3.9 Resistance to fluoroquinolones

Resistance to fluoroquinolones arises by mutation in the DNA gyrase or topoisomerase 1 V coding bacterial chromosome gene or by successful drug transport out of the cell [73]. Topoisomerase 1 V mutations can occur in gyrA / gyrB genes within the motif of the quinolone-resistant determinative region (QRDR), which is considered to be the active site of the enzyme. This contributes to the altered amino acid sequences of the subunits A and B, and hence to the altered topoisomerase II with a low affinity for quinolone molecules. As a result of point mutations in parC and parE genes encoding the ParC and ParE enzyme subunits, modifications of a secondary target (topoisomerase IV) occur. The over-expression of efflux includes other types of fluoroquinolone tolerance in *Pseudomonas*. Mutations in the nalB, nfxB and nfxC genes, resulting in overexpression of MexA-MexB-OprM, MexC-MexD- OprJ and MexE- MexF- OprN fallowing efflux [74].

3.10 Biofilm-mediated resistance

A biofilm is an aggregate of microorganisms that bind to each other on a living or non-living surface and are embedded in an extracellular polymeric (EPS) matrix of self-produced substances, including exopolysaccharides, proteins, metabolites, and eDNA [75, 76]. The microbial cells grown in biofilms are less sensitive than the cells grown in free aqueous suspension to the antimicrobial agents and the host immune response [77]. Even bacteria that are deficient or lack protective mutations in their intrinsic resistance, when they grow in a biofilm, they can become less susceptible to antibiotics [78]. The general mechanisms of biofilm-mediated resistance that protect bacteria from antibiotic attack include antibiotic penetration prevention, altered microenvironment that induces slow biofilm cell growth, adaptive stress response induction, and differentiation of persistent cells [78–80].

P. aeruginosa causes chronic lung infections in CF patients and, through the production of DNA, proteins and exopolysaccharides, forms a biofilm on lung epithelial cell surfaces. The regulation of the formation of *P. aeruginosa* biofilm is multifactorial and mainly depends on quorum sensing systems, GacS / GacA and RetS / LadS two-component regulatory systems, exopolysaccharides and cdi- GMP [81]. Quorum sensing is a form of communication between bacterial cells and cells that regulates gene expression in response to changes in cell population density. *P. aeruginosa* has three major systems of quorum sensing, LasILasR, RhlI-RhlR, and PQS-MvfR, all of which contribute to mature and differentiated biofilm formation. During biofilm formation, *P. aeruginosa* undergoes numerous physiological and phenotypic changes [82]. For example, *P. aeruginosa* strains convert to a mucoid phenotype in CF chronic infection that displays upregulated production of alginate driven by the CF microenvironment, enabling the formation of colonies of biofilms. Due to its ability to show swarming and twitching motility, *P. aeruginosa* flagellum is important for the initiation of biofilm formation. However, *P. aeruginosa* significantly decreases flagellum expression after surface attachment and may also permanently lose the flagellum due to genetic mutations, reducing host immune response activation, allowing *P. aeruginosa* to evade immune detection and phagocytosis [83].

4. The global economic scenario of antibiotic resistance

It is still an immense global challenge to quantify the exact economic effect of resistant bacterial infections. Measuring the distribution of the disease associated with antibiotic resistance is a crucial prerequisite in this situation. A major economic burden for the entire world is antibiotic resistance. In the USA alone, 99,000 deaths are caused annually by antibiotic-resistant pathogen-associated hospital-acquired infections (HAIs). Approximately 50,000 Americans died in 2006 because of two popular HAIs, namely pneumonia and sepsis, costing the US economy around $8 billion [84]. Patients with antibiotic-resistant bacterial infections need to remain in the hospital for at least 13 days, creating an extra 8 million hospital days each year. There have been estimates of costs of up to $29,000 per patient infected for an antibiotic-resistant bacterial infection. In total, economic losses of approximately $ 20 billion were recorded in the US, while losses of approximately $35 billion per year were also recorded in terms of loss of productivity due to antibiotic resistance in health care systems [85].

A worst-case scenario could emerge in the coming future, according to the analysts of the Research and Development Corporation, a US non-profit global organization, where the planet could be left without any effective antimicrobial agent to treat bacterial infections. In this case, the global economic burden will be nearly $120 trillion ($3 trillion per annum), roughly equal to the entire actual annual health care budget of the United States. In general, the world population will be significantly affected: about 444 million people will succumb to infections as of 2050, and birth rates will decrease rapidly in this scenario [86, 87]. These losses are calamitous, but these estimates reflect imperfect images of the economic costs of antibiotic resistance due to data limitations such as the inclusion of total conditions and resistance-susceptible diseases. The use of antibiotics in the livestock and food industries is another very critical trait of antimicrobial resistance (AMR), that was missing from the investigation. It is an important player in the rising AMR, likely causing its own projected economic losses. There is also a misappropriation of the use of antimicrobials as growth promoters in many developing countries. This activity has been outlawed in the European Union since 2006 [88, 89].

5. Novel alternative antimicrobial therapy for *P. aeruginosa* treatment

The overuse and misuse of antibiotics, which can lead to unwanted side effects and the production of drug-resistant bacterial strains, is a growing public health issue. The production of new antibiotics, in addition, is very limited and timely. The development of innovative therapeutic approaches to the treatment of infections with *P. aeruginosa* is therefore highly desirable and has received further interest over the past decade. These innovative therapeutic techniques, which involve antimicrobial peptides, phage therapy, inhibition of quorum sensing as well as the use of iron chelation, nanoparticles, probiotics and vaccine strategies.

5.1 Antimicrobial peptides

A number of species, from bacteria to animals, develop antimicrobial peptides (AMPs), also called host defense peptides, and they are active against a wide range of microorganisms [90]. There is no complete understanding of the mode(s) of operation of AMPs. It is widely agreed that the cytoplasmic membrane is attacked by AMPs, leading to cell death [91]. AMPs have also been shown to possess anti-biofilm and immunomodulatory properties, in addition to antimicrobial activity, AMPs have been proposed as an alternative to traditional antibiotics to battle bacterial infections as a result of their broad-spectrum activity; AMPs exhibit rapid killing kinetics, low mediated resistance levels, and low host toxicity [92]. Many antimicrobial peptides, including GL13 K, LL-37, T9W, NLF20, cecropin P1, indolicidin, magainin II, nisin, ranalexin, melittin, and defensin, have demonstrated powerful antimicrobial effects of either direct bactericidal effects or biofilm disruption against *P. aeruginosa* [93]. In addition, by facilitating antibiotic absorption, disrupting biofilm formation or inhibiting bacterial quorum, some AMPs have demonstrated synergy with traditional antibiotics against several bacteria, including *P. aeruginosa* [94]. For instance, it has been shown that the clearance of *P. aeruginosa* biofilm was increased by a combination of GL13 K with tobramycin [95]. In 2017, Zheng et al. [96] observed that when combined with tetracycline in vitro, the minimum inhibitory concentration of cecropin A2 against clinical isolates of *P. aeruginosa* was reduced 8-fold.

5.2 Phage therapy

By inducing lysis, bacteriophages (phages) are viruses that infect and destroy bacteria [97]. In 1915, the British bacteriologist Frederick Twort first discovered phages. Two years later, Félix d'Herelle made a similar discovery independently in Paris and presented the phage therapy notion. With the advent of antibiotic therapy, phage therapy was abandoned in several countries, but has been continuously developed in Eastern European countries with facilities in Warsaw, Poland, and Tbilisi, Georgia [98]. Shotgun metagenome sequencing showed that there were antipseudomonal phages in the phage cocktails sold in pharmacies in Georgia and Russia [99]. The successful treatment of infections with MDR *P. aeruginosa* has been reported in a few case reports from Belgium and the US, but has not gained broad acceptance in the Western world [100]. Phage therapy has many benefits, including replication at the infection site, high precision for attacking bacteria without effects on commensal flora, less side effects than other therapies, antibiotic-resistant bacteria bactericidal activity and simple administration [101]. The use of phages as an alternative to antibiotics has been extensively studied for the treatment of *P. aeruginosa* infections. There are 137 different phages that have been characterized to date that target the *Pseudomonas* genus [102]. Many in vitro and in vivo studies have been performed to test the efficacy of phages against chronic infections of *P. aeruginosa*. For instance, co-incubation of phage PA709 with the clinical strain *P. aeruginosa* 709 has been shown to significantly reduce the viability of *P. aeruginosa*. Another research found that intranasal administration of P3-CHA bacteriophage to mice receiving a lethal dose of *P. aeruginosa* strain CHA substantially improved the rate of survival and reduced the bacterial load in the lungs [103].

Another benefit of phage therapy is that phages can be genetically modified as vehicles to transport bacteria with antimicrobial agents, thus increasing treatment efficacy [104]. While phages have been shown to be successful in vitro and in animal models against bacterial infection, only a small number of phage therapy clinical trials have been performed to date. The reasons for this include: safety issues about post-treatment phage clearance and impurity of phage preparations; poor stability of phage preparations; and lack of knowledge of the comprehensive phage mode of action and bacterial resistance to phage growth [105]. In clinical trials, the use of phages against *P. aeruginosa* infections has been studied in patients with venous leg ulcers, burn wounds and otitis, and no adverse reactions have been identified during these clinical trials [106].

5.3 Quorum sensing inhibition

Quorum sensing is a mechanism that enables bacteria to regulate the expression of genes in a manner based on cell density. To control virulence and biofilm formation, *P. aeruginosa* utilizes quorum sensing [107]. Las and Rhl are two major *P. aeruginosa* quorum-sensing systems responsible for the synthesis of the signal molecules of N-acyl homoserine lactone (AHL), N-(3-oxododecanoyl)-L-homoserine lactone (3O-C12-HSL) and N-butanoyl-L-homoserine lactone (C4-HSL). 3O-C12-HSL and C4-HSL bind to and activate their LasR and RhlR cognate transcription factors, respectively, inducing the formation of biofilms and the expression of various virulence factors, including elastase, proteases, pyocyanin, lectins, rhamnolipids, and toxins [108]. The third *P. aeruginosa* quorum-sensing system, PQSMvfR, has been reported to facilitate the formation of biofilms in addition to the LasI-LasR and RhlI-RhlR systems. This mechanism regulates the development of the *Pseudomonas* quinolone signal (PQS), 2-heptyl-3-hydroxy-4-quinolone, by the transcriptional regulator MvfR, also known as PqsR, by controlling the pqsABCDE operon. In addition, PqsA and PqsD proteins have been implicated in the development of biofilms [82].

A promising technique for treating *P. aeruginosa* infections is known to be the inhibition of quorum sensing. This approach is capable of preventing or decreasing the formation of biofilms, reducing bacterial virulence and has a low risk of bacterial resistance growth. In addition, this strategy has a small scope, such that any unwanted inhibitory effects on beneficial bacteria are impossible. For the Las and Rhl systems, quorum sensing inhibitors may be either natural or synthetic and are capable of reducing the activity of AHL synthase, inhibiting the development of AHL, degrading AHLs or competing for AHL receptor binding [109]. In recent years, the use of quorum sensing inhibitors for the treatment of infections with *P. aeruginosa* has been intensively studied. The carotenoid zeaxanthin, typically found in plants, algae and lichens, for example, reduced the formation of biofilms in *P. aeruginosa* by binding to the signal receptors for quorum sensing, lasR and RhlR, and blocking the expression of virulence genes, lasB and rhlA [110]. Flavonoids are a class of naturally developed plant metabolites that have acted as LasR and RhlR antagonists and substantially decreased their ability to bind to the *P. aeruginosa* promoters of quorum sensing-regulated genes [111].

5.4 Iron chelation

Iron is important for bacterial growth and is involved in a number of cellular processes, such as the production of electricity, the replication of DNA and the transport of electrons [112]. Compared to healthy people, the iron content of human sputum was found to be substantially elevated in CF patients, indicating that an increased amount of iron promotes chronic CF lung infection [113]. *P. aeruginosa* utilizes pyoverdine and pyochelin siderophores to obtain iron from the extracellular environment [114]. Therefore, a technique to fight *P. aeruginosa* infections is to limit the concentration of extracellular iron or disrupt iron uptake by *P. aeruginosa*. Several studies have related iron metabolism to the pathogenesis of chronic infections, indicating that iron analogues and chelators may work against *P. aeruginosa* as potential therapeutic agents. For example, iron chelators, 2,2′-dipyridyl (2DP), diethylenetriaminepentacetic acid (DTPA) and EDTA, have been reported to impair growth and biofilm formation of *P. aeruginosa* and have been more effective under anaerobic conditions [115].

Gallium is a nonredox iron III analog that disrupts the metabolism of bacterial iron by acting in several biological processes as an iron replacement, so it is a US

FDA-approved medication for cancer-associated hypercalcemia treatment [116]. In 2007, Kaneko et al., reported that gallium was able to inhibit the growth of *P. aeruginosa*, prevent the development of biofilms, and manifest excellent bactericidal activity in vitro by reducing the uptake of bacterial iron and repressing the synthesis of pyoverdine mediated by the transcriptional regulator PvdS. In addition, in mouse infection models, gallium has also been found to remove *P. aeruginosa* effectively.

5.5 Nanoparticles

Currently, a variety of diseases, including cancer and bacterial infectious diseases, have received significant attention from nanoparticles to treat them. Nanoparticles are small materials that have been used in a number of chemical, biological and biomedical applications, having a size of less than 100 nm and a large surface area to mass ratio [117]. The nanoparticles used for their antimicrobial activity are highly penetrable in the bacterial membranes, may interfere with the formation of biofilms, have several antimicrobial mechanisms, and are strong antibiotic carriers [118]. For the prevention of *P. aeruginosa* infections, metallic and antimicrobial agent-loaded nanoparticles have been extensively examined. Silver nanoparticles, for example, are powerful antimicrobial agents that generate silver ions responsible for the inhibition, like DNA synthesis, of bacterial enzymatic systems. Silver nanoparticles have shown important antimicrobial effects on the clinical strains of *P. aeruginosa*, killing *P. aeruginosa* effectively and inhibiting its in vitro growth. In addition, silver nanoparticles have demonstrated low mammalian cell cytotoxicity, although this requires more in vivo research [119].

Nanoparticles are capable of delivering antimicrobial agents such as antibiotics to bacteria, as described earlier. Kwon et al., developed porous silicon nanoparticles with a novel antimicrobial peptide fused with a synthetic bacterial toxin, containing membrane-interacting peptides. This engineered nanoparticle was discovered in a mouse model of *P. aeruginosa* lung infection to increase the survival rate and bacterial clearance. Moreover, it has been found that the binding of antibiotics to nanoparticle surfaces greatly improves the effectiveness of both antibiotics and nanoparticles. In this respect, silver ampicillin-attached nanoparticles have a higher rate of in vitro killing of ampicillin-resistant *P. aeruginosa* isolates compared to silver ampicillin-attached nanoparticles [120].

5.6 Probiotic as an alternative antimicrobial therapy

Probiotics are living microorganisms which, when ingested in appropriate quantities, provide health benefits [121]. The majority of probiotic bacteria are gram-positive, and their primary functions are related to intestinal tract health regulation and maintenance (e.g., *Lactobacillus* and *Bifidobacterium*) [122]. The probiotics in the intestines that colonize the human host are the most numerous. The commensal intestinal microbiome leads to enhanced infection tolerance, differentiation of the host immune system, and nutrient synthesis [123]. The probiotic *Pediococcus acidilactici* HW01 was studied against *P. aeruginosa* and observed decreased *P. aeruginosa* motility as well as decreased pyocyanin development, decreased protease and rhamnolipid production, and decreased stainless steel surface biofilm formation. Another research conducted by Moraes et al., showed that *Lactobacillus brevis* and *Bifidobacterium bifidum* were effective against *S. aureus* biofilms grown on titanium discs. The findings showed a decrease in *S. aureus* growth on titanium discs when both probiotics were used, but *L. brevis* strains was shown to have the greatest inhibitory effect on biofilm formation.

Recent studies by Xu et al. [124], have indicated that probiotics can be used by patients infected by COVID 19 to prevent secondary infections. There was intestinal microbial dysbiosis in some patients with COVID-19. In all patients, nutritional and gastrointestinal functions must be measured. To control the composition of the intestinal microbiota and reduce the risk of secondary infection due to bacterial translocation, nutritional support and application of probiotics was suggested.

5.7 Vaccine strategy

The concept of a vaccine strategy is to avoid infection until it can be produced. The production of vaccines aims to prevent and decrease infections of *P. aeruginosa* [125]. However, no approved vaccine against this pathogen is yet available. *P. aeruginosa* antigens, which are responsible for pathogenesis, induce potent immune responses. LPS O-antigen, polysaccharide protein conjugates, outer membrane proteins OprF and OprI, type III secretion system portion PcrV, flagella, pili, DNA, live-attenuated *P. aeruginosa* and whole killed cells are possible candidates for *P. aeruginosa* vaccines [126]. Among the potential *P. aeruginosa* vaccines, phase III clinical trials in CF patients were performed only with the flagella vaccine and the recombinant vaccine IC43, containing OprF and OprI.

Related to the ability of this pathogen to undergo phenotypic changes in variable environmental conditions, the existing vaccines for *P. aeruginosa* demonstrate poor efficacy in clinical trials. For example, *P. aeruginosa* strains downregulate the expression of highly immunogenic virulence factors in CF patients' lungs, such as LPS O-antigen, type III secretion systems, flagella and pili [127]. In addition, impaired mechanisms of host protection often reduce vaccination effectiveness. Due to the CF lungs having an altered mucus layer, impaired phagocytosis, and dysregulated inflammatory responses, including aberrant cytokine and chemokine production, and reduced phagocyte recruitment, the lung microenvironment in CF patients has become a great challenge for effective vaccination [128].

6. Role of combination therapy *versus* monotherapy

Early administration of adequate antibiotic therapy was associated with a favorable clinical outcome, especially among critically ill patients with serious *P. aeruginosa* infections [129]; on the other hand, delays in administering adequate antibiotic therapy were associated with a substantial increase in mortality. The progressive rise in antibiotic resistance in *P. aeruginosa* has been established in recent years as the key explanation for the inadequacy of antibiotics, with a negative effect on patient survival [130].

The evidence available indicates that the greatest advantage of combination therapy derives from an increased probability of selecting an appropriate agent during empirical therapy, rather than avoiding resistance during definitive therapy or benefiting from synergistic action in vitro. Therefore, researchers recommend early administration of a combination regimen when *P. aeruginosa* is suspected, followed by a prompt de-escalation when the antimicrobial susceptibility test becomes available, to balance between early antibiotic administration and the risk of resistance selection. An approach consisting of the prescription of an anti-pseudomonal beta-lactam (piperacillin / tazobactam, ceftolozane / tazobactam, ceftazidime, cefepime, or carbapenem) plus a second (aminoglycoside or fluoroquinolone) anti-pseudomonal agent is encouraged.

7. New antipseudomonal antibiotics

Related to the emergence of multidrug-resistant strains, traditional antibiotic therapies against *P. aeruginosa* infections have become increasingly ineffective. The use of various antibiotic combinations and the development of new antibiotics are existing therapeutic options for *P. aeruginosa* treatment. New antibiotics have been shown to be more effective in destroying *P. aeruginosa* and have a lower frequency of production of resistance compared to current antibiotics due to their novel modes of action, efficient delivery of drugs (e.g. inhaled antibiotics) and resistance to bacterial enzyme alteration. Novel antibiotics with action against *P. aeruginosa* have been available in Europe in recent years and others are in advanced stages of clinical development. In certain instances, indirect evidence indicates their possible superiority over standard anti-pseudomonal regimes.

7.1 Doripenem

Doripenem is a new carbapenem antibiotic with wide spectrum activity against gram-negative and gram-positive bacteria by binding to penicillin-binding proteins by inhibiting bacterial cell wall synthesis; it has been approved for the treatment of complicated intra-abdominal infection and urinary tract infection by the US Food and Drug Administration (FDA) [131]. Except for the metallo-β-lactamases of class B, doripenem is resistant to hydrolysis by several β-lactamases. Importantly, compared to other carbapenem antibiotics such as meropenem and imipenem, the in vitro antibacterial activity of doripenem against the *P. aeruginosa* isolates from CF patients was found to be more active [132]. In addition, the effectiveness of doripenem was tested in patients with *P. aeruginosa* ventilator-associated pneumonia, a phase III clinical trial of patients with *P. aeruginosa* ventilator-associated pneumonia found that patients treated with doripenem had higher rates of cure compared to patients treated with imipenem. Of note, headache, nausea, diarrhea, rash, and phlebitis are among the side effects of doripenem [133].

7.2 Plazomicin

Plazomicin is a semisynthetic aminoglycoside antibiotic of the next generation that is synthetically derived from the natural product sisomicin. A wide range of aminoglycoside modifying enzymes, but not 16S rRNA ribosomal methyltransferases, are able to resist plazomicin [134]. Plazomicin exhibits potent in vitro activity against both gram-negative and gram-positive bacterial pathogens and has an activity close to that of amikacin against strains of multidrug-resistant *P. aeruginosa*. In addition, Pankuch et al., reported in vitro synergistic activity of plazomicin against clinical isolates of *P. aeruginosa* in combination with cefepime, doripenem, imipenem or piperacillin-tazobactam and no antagonism was observed in this study, indicating that plazonmicin is a possible candidate for combination therapy in the treatment of infections with multidrug-resistant *P. aeruginosa*. Plazomicin can cause nephrotoxic and ototoxic effects that are mild to moderate [135].

7.3 POL7001

As a novel class of antibiotics against *P. aeruginosa*, protein epitope mimetic (PEM) molecules have emerged; some PEM molecules inhibit the transfer of LPS to the outer bacterial membrane [136]. A macrocycle molecule belonging to the PEM antibiotic family is POL7001. The efficacy of POL7001 was tested by Cigana et al., both in vitro and in murine *P. aeruginosa* acute and chronic pneumonia models. They observed that *P. aeruginosa* multidrug-resistant isolates were susceptible to POL7001 in CF patients, and that POL7001-treated mice had substantially decreased bacterial burden and decreased levels of lung inflammation during acute and chronic *P. aeruginosa* infection. POL7001 as a novel therapeutic agent for potential clinical trials is indicated by the new mode of action, effective pulmonary delivery and potent in vitro and in vivo activity. The side effects of POL7001 have not yet been identified [137].

7.4 Arikayce ™

Arikayce™ has been approved by the FDA for the treatment of *Mycobacterium avium* complex (MAC) lung disease, and is a liposomal amikacin treatment. Clinical trials for this drug and its efficacy in the treatment of *P. aeruginosa* in patients with cystic fibrosis have been performed. While these are early phases and some experiments would have to resolve the drawbacks of this compound in order to improve safety, some experimental clinical trials have been performed [138].

7.5 Ceftolozane-tazobactam

To resolve *P. aeruginosa* antimicrobial resistance mechanisms, such as changes in porine permeability and upregulation of efflux pumps, ceftolozane-tazobactam is being created. Due to a higher affinity for all essential PBPs, including PBP1b, PBP1c and PBP3, this drug has an intrinsically potent anti-pseudomonal effect [139]. Ceftolozane/ Tazobactam has been shown to have a strong in vitro activity against most strains of MDR *P. aeruginosa* [including strains developing extended-spectrum β-lactamase (ESBL) but not carbapenemase]. The therapeutic use of ceftolozane-tazobactam in complicated intra-abdominal and urinary tract infections has been suggested by the FDA [140]. In addition, a study is currently underway for the treatment of HAP, including ventilator-associated pneumonia (VAP). In 71 percent of patients with MDR *P. aeruginosa* infections, evidence from real-world trials using ceftolozane-tazobactam for the treatment of MDR *P. aeruginosa* infections showed promising results.

7.6 Ceftazidime-avibactam

Ceftazidime-avibactam is a novel combination of β-lactam / BLI approved for the treatment of complicated urinary tract infections (UTIs) and complicated intra-abdominal infections by the Food and Drug Administration (FDA) and *European Medicines Agency* (EMA). In vitro studies have shown that the combination of ceftazidime-avibactam is highly effective against KPC- producing *Klebsiella pneumoniae* carbapenemase (KPCs), oxacillinase (OXA), extended- spectrum β-lactamases (ESBLs) and AmpC enzymes producing Enterobacteriaceae. The drug does not, however, have any action against metallo-beta β- lactamases (MBL, VIM and NDM) and avibactam does not have any improved activity against *P. aeruginosa* [141]. In phase III research compared ceftazidime-avibactam to meropenem (NTC01808092), the efficacy of ceftazidime-avibactam against VAP was analyzed [142]. The predominant isolated baseline gram-negative pathogens were *K. pneumoniae* and *P. aeruginosa*, with 28% of patients possessing a non-susceptible isolate of 1% ceftazidime. 356 patients in the clinically evaluable population were treated with ceftazidime-avibactam and 370 with meropenem. The research met the non-inferiority criterion for ceftazidime-avibactam as there was no disparity in the outcome between the groups. In addition, the efficacy of ceftazidime-avibactam was close to that of ceftazidime-susceptible pathogens against ceftazidime-non-susceptible strains and was also comparable to meropenem.

7.7 Imipenem-cilastatin-relebactam

Relebactam is a β-lactamase inhibitor (BLI) diazabicyclooctane that inhibits β-lactamase class A and C activity, but has no activity against metallo-β-lactamase. It has been shown that the combination of imipenem-cilastatin with relebactam has synergistic activity against a broad range of MDR gram negative pathogens including *P. aeruginosa*, KPC-producing *K. pneumoniae and Enterobacter* spp. [143]. This medication has been tested predominantly in patients with IAI, complicated UTI, and pyelonephritis, although a trial is currently underway in patients with HAP/VAP. Some new medications have a small effect on *P. aeruginosa*, such as plazomycin, meropenem-vaborbactam and aztreonam-avibactam [144].

8. Conclusions

Treatment of infections with *P. aeruginosa* continues to be significant challenging. Improving the early diagnosis and empirical treatment of serious *P. aeruginosa* infections is an urgent need. First, to quickly announce the detection and susceptibility results for *Pseudomonas* in blood cultures and other clinically important cultures, matrix-assisted-laser-desorption-ionization time-of-flight mass spectrometry (MALDI-TOF) and modern molecular techniques should be routinely introduced. However, in order to decide if such diagnostic methods have a real effect on hospitalization time and patient mortality, controlled trials would be required. Secondly, more studies are urgently required to classify patients at risk of infection with MDR *P. aeruginosa* (bloodstream infections, urinary tract infections) based on clinical risk factors. Ultimately, clinical response depends on factors such as underlying diseases, seriousness of infection, form of infection, adequate control of the source, and response to prior antibiotics. There is an immediate need to determine the true impact of the latest anti-Pseudomonas drugs recently approved for the treatment of these infections on patient outcomes. To date, however, due to high cost, side effects and safety issues, few of these newer methods have continued to clinical practice.

Author details

Rahman Laibi Chelab
Faculty of Education for Pure Science, Thi-Qar University, Thi-Qar, Iraq

*Address all correspondence to: rahmanlaeby2007@yahoo.com

References

[1] Olga P, Lucia B, Inès B, Laura FG, Mónica GB, Antón A, María L, German B, Maria T. 2020. Strategies to Combat Multidrug-Resistant and Persistent Infectious Diseases. Antibiotics.2020; 9, 65. DOI: 10.3390/antibiotics9020065

[2] WHO. Worldwide Country Situation Analysis: Response to Antimicrobial Resistance; WHO Library Cataloguing-in-Publication Data;World Health Organization: Geneva, Switzerland, 2015. ISBN: 978 92 4156494 6

[3] Cerceo E, Deitelzweig SB, Sherman BM, Amin AN. Multidrug-Resistant Gram-Negative Bacterial Infections in the Hospital Setting: Overview, Implications for Clinical Practice, and Emerging Treatment Options. Microbial Drug Resistance. 2016;22, 412-431. DOI: 10.1089/mdr.2015.0220

[4] El Solh AA, Alhajhusain A. Update on the treatment of *Pseudomonas aeruginosa* pneumonia. Antimicrobial Agents and Chemotherapy. 2009; 64, 229-238. DOI: 10.1093/jac/dkp201

[5] Park SY, Park HJ, Moon SM, Park KH, Chong YP. Impact of adequate empirical combination therapy on mortality from bacteremic *Pseudomonas aeruginosa* pneumonia. BMC Infectious Diseases. 2012; 12, 308. DOI: 10.1186/1471-2334-12-308

[6] Lister PD, Wolter DJ, Hanson ND. Antibacterial-resistant *Pseudomonas aeruginosa:* clinical impact and complex regulation of chromosomally encoded resistance mechanisms. Clinical Microbiology Reviews. 2009; 22, 582-610.DOI: 10.1128/CMR.00040-09

[7] Tacconelli, E., Magrini, N., Carmeli, Y., Harbarth, S., Kahlmeter, G., Kluytmans, J., Mendelson, M., Pulcini, C., Singh, N., Theuretzbacher, U. Global priority list of antibiotic-resistant bacteria to guide research, discovery, and development of new antibiotics. World Health Organization 1-7

[8] Hirsch EB, Tam VH. Impact of multidrug-resistant *P. aeruginosa* infection on patient outcomes. Expert Review of Pharmacoeconomics and Outcomes Research. 2010;10, 441-451. DOI: 10.1586/erp.10.49.

[9] Hancock RE, Speert DP. Antibiotic resistance in *Pseudomonas aeruginosa:* mechanisms and impact on treatment. Drug Resistance Updates. 2000; 3, 247-255. DOI: 10.1054/drup.2000.0152

[10] Breidenstein EB, Fuente-Nunez C, Hancock RE. *Pseudomonas aeruginosa*: all roads lead to resistance. Trends in Microbiology. 2011; 19, 419-426. DOI.10.1016/j.tim.2011.04.005

[11] Drenkard E. Antimicrobial resistance of *Pseudomonas aeruginosa* biofilms. Microbes and Infection. 2003; 5, 1213-1219. DOI: 10.1016/j.micinf.2003.08.009

[12] Burkhard TU. Emerging therapies against infections with *Pseudomonas aeruginosa*. 2019; 8(F1000 Faculty Rev):1371 . DOI: 10.12688/f1000research.19509.1

[13] Gorr SU, Flory CM, Schumacher RJ. In vivo activity and low toxicity of the second-generation antimicrobial peptide DGL13K. PLoS One. 2019; 14(5). DOI: 10.1371/journal.pone.0216669

[14] Yasir M, Dutta D, Willcox MDP. Comparative mode of action of the antimicrobial peptide melimine and its derivative Mel4 against *Pseudomonas aeruginosa*. Scientific Reports. 2019; 9(1): 7063. DOI: 10.1038/s41598-019-42440-2

[15] Romoli O, Mukherjee S, Mohid SA. Enhanced silkworm cecropin B antimicrobial activity against *Pseudomonas aeruginosa* from single amino acid variation. ACS Infectious Diseases. 2019; 5(7): 1200-13. DOI: 10.1021/asinfecdis. 9b00042

[16] Molchanova N, Wang H, Hansen PR. Antimicrobial activity of α- peptide/β-peptoid lysine-based peptidomimetics against colistin-resistant *Pseudomonas aeruginosa* isolated from Cystic Fibrosis patients. Frontiers of Microbiology. 2019; 10: 275. DOI: 10.3389/fmicb.2019.00275

[17] Candido ES, Cardoso MH, Chan LY. Short cationic peptide derived from archaea with dual antibacterial properties and anti-Infective potential. ACS infectious diseases. 2019; 5(7): 1081-1086. DOI: 10.1021/asinfecdis. 9b00073

[18] Kang HK, Seo CH, Luchian T. Pse-T2, an antimicrobial peptide with high-level, broad-spectrum antimicrobial potency and skin biocompatibility against multidrug-resistant *P. aeruginosa* infection. Antimicrobial Agents and Chemotherapy. 2018; 62(12). DOI: 10. 1128/ AAC. 01493-18

[19] Beaudoin T, Stone TA, Glibowicka M. Activity of a novel antimicrobial peptide against *P. aeruginosa* biofilms. Scientific Reports. 2018; 8(1): 14728. DOI: 10.1038/ s41598-018-33016-7

[20] Akbari R, Hakemi Vala M, Hashemi A. Action mechanism of melittinderived antimicrobial peptides, MDP1 and MDP2, de novo designed against multidrug resistant bacteria. Amino Acids. 2018; 50(9): 1231-43. DOI: 10.1007/s00726-018-2596-5

[21] Mulcahy LR, Burns JL, Lory S, Lewis K. Emergence of *Pseudomonas aeruginosa* strains producing high levels of persister cells in patients with cystic fibrosis. Journal of Bacteriology. 2010 D;192(23):6191-9. DOI: 10.1128/ JB.01651-09

[22] Walkty A, Adam H, Baxter M, Denisuik A, Lagace WP. In vitro activity of plazomicin against 5,015 gram-negative and gram-positive clinical isolates obtained from patients in canadian hospitals as part of the CANWARD study, 2011-2012. Antimicrobial Agents and Chemotherapy. 2014; 58(5): 2554-63. DOI: 10.1128/AAC.02744-13

[23] Cigana C, Bernardini F, Facchini M, Alcala-Franco B, Riva C. Efficacy of the Novel Antibiotic POL7001 in Preclinical Models of *Pseudomonas aeruginosa* Pneumonia. Antimicrobial Agents and Chemotherapy. 2016; 60, 4991-5000. DOI: 10.1128/AAC.00390-16

[24] Chatterjee M, Anju CP, Biswas L, Anil Kumar V, Gopi Mohan C, Biswas R. Antibiotic resistance in *Pseudomonas aeruginosa* and alternative therapeutic options. International Journal of Medical Microbiology. 2016;306(1):48-58. DOI: 10.1016/j.ijmm.2015.11.004.

[25] Guilhelmelli F, Vilela N, Albuquerque P, Derengowski LS, Silva-Pereira S, Kyaw CM. Antibiotic development challenges: the various mechanisms of action of antimicrobial peptides and of bacterial resistance. Frontiers of Microbiology. 2013;4:353. DOI: 10.3389/fmicb.2013.00353

[26] Hooper DC. Mechanisms of Action of Antimicrobials: Focus on Fluoroquinolones. Clinical Infectious Diseases. 2001;32 (Suppl 1):S9-S15. DOI: 10.1086/319370.

[27] Kotra LP, Mobashery S. Beta-lactam antibiotics, Beta-lactamases and bacterial resistance. Bulletin of Institute Pasteur. 1998; 96: 139-150. DOI: 10.1016/S0020-2452(98)80009-2

[28] Wilke MS, Lovering AL, Strynadka NCJ. Beta-Lactam antibiotic

resistance: a current structural perspective. Current Opinion in Microbiology. 2005;8:525-533. DOI: 10.1016/j.mib.2005.08.016

[29] Neu HC, Gootz TD. Baron S. Medical Microbiology. 4th edition. Galveston (TX): University of Texas Medical Branch at Galveston; 1996. ISBN-10: 0-9631172-1-1

[30] Peterson JW, Baron S. Bacterial Pathogenesis. Medical Microbiology. 4th edition. Galveston (TX): University of Texas Medical Branch at Galveston; 1996. DOI: 10: 0-9631172-1-1

[31] Krupovi_CM, Daugelavi_cius R, Bamford DH. Polymyxin B Induces Lysis of Marine Pseudoalteromonads. Antimicrobial Agents and Chemotherapy. 2007; 51(11):3908-3914. DOI: 10.1128/aac.00449-07

[32] McKee EE, Ferguson M, Bentley AT, Marks TA. Inhibition of mammalian mitochondrial protein synthesis by oxazolidinones. Antimicrobial Agents and Chemotherapy. 2006;50(6):2042-9. DOI: 10.1128/AAC.01411-05

[33] Cocito C, Di Giambattista M, Nyssen E, Vannuffel P. Inhibition of protein synthesis by streptogramins and related antibiotics. Journal of Antimicrobial Chemotherapy. 1997; 39:7-13. DOI:10.1093/JAC/39.SUPPL-1.7

[34] Kohanski MA, Dwyer DJ, Collins JJ. How antibiotics kill bacteria: from targets to networks. Nature Reviews Microbiology, 2010; 8:423-435. DOI: 10.1038/nrmicro2333

[35] Liu LF, Liu CC, Alberts BM. Type II DNA topoisomerases: enzymes that can unknot a topologically knotted DNA molecule via a reversible double-strand break. Cell. 1980;19(3):697-707. DOI: 10.1016/s0092-8674(80)80046-8

[36] Ehmann DE, Lahiri SD. Novel compounds targeting bacterial DNA topoisomerase/DNA gyrase. Current Opinion in Pharmacology. 2014 ;18:76-83. DOI: 10.1016/j. coph. 2014.09.007

[37] Chatterji, Unniraman MS, Mahadevan S, Nagaraja V. Effect of different classes of inhibitors on DNA gyrase from *Mycobacterium smegmatis*. Journal of Antimicrobial Chemotherapy. 2001;48(4):479-485. DOI: 10.1093/jac/48.4.479

[38] Collin F, Karkare S, Maxwell A. Exploiting bacterial DNA gyrase as a drug target: current state and perspectives. Applied Microbiology and Biotechnology. 2011; 92(3):479-497. DOI: 10. 1007/s00253-011-3557-z

[39] Bagge N, Hentzer M, Andersen JB, Ciofu O, Givskov M, Hoiby N. Dynamics and spatial distribution of Beta-lactamase expression in *Pseudomonas aeruginosa* biofilms. Antimicrobial Agents and Chemotherapy. 2004; 48(4): 1168-1174. DOI: 10.1128/aac.48.4. 1168-1174.2004

[40] Juan C, Macia MD, Gutierrez O, Vidal C, Perez JL, Oliver A. Molecular mechanisms of beta-lactam resistance mediated by AmpC hyper production in *P. aeruginosa* clinical strains. Antimicrobial Agents and Chemotherapy. 2005;49(11):4733-4738. DOI: 10.1128/AAC.49.11.4733-4738.2005

[41] Bush K, Jacoby GA, Medeiros AA. A functional classification scheme for beta-lactamases and its correlation with molecular structure. Antimicrobial Agents and Chemotherapy. 1995;39(6):1211-33. DOI: 10.1128/aac.39.6.1211

[42] Martin-Loeches I, Deja M, Koulenti D. Potentially resistant microorganisms in intubated patients with hospital acquired pneumonia: the interaction of ecology, shock and risk factors. Intensive Care Medicine. 2013;39(4):672-681. DOI: 10.1007/s00134-012-2808-5

[43] Okomoto K, Gotoh N, Nishino T. *Pseudomonas aeruginosa* reveals high intrinsic resistance to penem antibiotics: penem resistance mechanisms and their interplay. Antimicrobial Agents and Chemotherapy. 2001; 45(7): 1964-1971. DOI: 10. 1128/AAC.45.7.1964-1971

[44] Paul M, Benuri-Silbiger I, Soares-Weiser K, Leibovici L. Beta lactam monotherapy versus beta lactam-aminoglycoside combination therapy for sepsis in immune competent patients: systematic review and meta-analysis of randomized trials. British medical journal. 2004; 328 (7441): 668. DOI: 10. 1136/ bmj. 38028.520995.63

[45] Poole K. Aminoglycoside resistance in *Pseudomonas aeruginosa*. Antimicrobial Agents and Chemotherapy. 2005;49(2):479-487. DOI: 10.1128/AAC.49.2.479-487

[46] Rello J, Allegri C, Rodriguez A. Risk factors for ventilator-associated pneumonia by *Pseudomonas aeruginosa* in presence of recent antibiotic exposure. Anesthesiology. 2006; 105(4):709-714.DOI: 10.1097/00000542-200610000-00016

[47] Horcajada JP, Shaw E, Padilla B. Healthcare-associated, community-acquired and hospital-acquired bacteraemic urinary tract infections in hospitalized patients: a prospective multicentre cohort study in the era of antimicrobial resistance. Clinical microbiology and infection. 2013;19(10):962-8. DOI: 10.1111/1469-0691.12089

[48] Honore N, Nicolas MH, Cole ST. Regulation of enterobacterial cephalosporinase production: the role of a membrane-bound sensory transducer. Molecular Microbiology. 1989; 3(8):1121-1130. DOI: 10.1111/j.1365-2958.1989.tb00262.x

[49] Kock R, Becker K, Cookson B. Methicillin-resistant *S. aureus* (MRSA): burden of disease and control challenges in Europe. Eurosurveillance. 2010;15(41):19688. DOI: 10.2807/ese.15.41.19688-en.

[50] Vakulenko SB, Mobashery S. Versatility of aminoglycosides and prospects for their future. Clinical Microbiology Reviews. 2003; 16(3): 430-450. DOI: 10.1128/ CMR.16.3.430-450.2003

[51] Ramirez MS, Tolmasky ME. Aminoglycoside modifying enzymes. Drug Resistace Updates. 2010; 13(6): 151-171. DOI: 10.1016/j.drup.2010.08.003

[52] Poole K. 2011. *Pseudomonas aeruginosa*: resistance to the max. Frontiers of Microbiology. 2011; 2: 65. DOI: 10.3389/fmicb.2011.00065

[53] Liano-Sotelo B, Azucena EF, Kotra LP, Mobashery S, Chow CS. Aminoglycosides modified by resistance enzymes display diminished binding to the bacterial ribosomal aminoacyl- tRNA site. Chemistry&Bio logy.2002;9(4):455-463 DOI: 10. 1016/ s1074-5521 (02) 00125-4

[54] Miller GH, Sabatelli FJ, Hare RS. The Aminoglycoside Resistance Study Group. The most frequent aminoglycoside resistance mechanisms e changes with time and geographic area: a reflection of aminoglycoside usage patterns. Clinical Infectious Diseases. 1997; 24 Suppl 1:S46-62. DOI: 10.1093/ clinids/24.supplement-1.s46

[55] MacLeod DL, Nelson LE, Shawar RM. Aminoglycoside resistance mechanisms for cystic fibrosis *Pseudomonas aeruginosa* isolates are unchanged by long-term, intermittent, inhaled tobramycin treatment. The Journal of Infectious Diseases. 2000;181(3):1180-1184. DOI: 10.1086/315312

[56] Yamane K, Doi Y, Yokoyama K. Genetic environments of the rmtA gene in *Pseudomonas aeruginosa* clinical

isolates. Antimicrobial Agents and Chemotherapy 2004; 48(6): 2069-2074. DOI: 10.1128/AAC.48.6.2069-2074.2004

[57] Gast Gurung M, Moon DC, Tamang MD. Emergence of 16S rRNA methylase gene armA and cocarriage of blaIMP-1 in *Pseudomonas aeruginosa* isolates from South Korea. Diagnostic Microbiology and Infectious Disease.2010;68:468-470. DOI: 10.1016/j.diagmicrobio.2010.07.021

[58] Hooper LV, Gordon JI. Commensal host-bacterial relationships in the gut. Science. 2001;292(5519):1115-1118. DOI: 10.1126/science.1058709

[59] Furtado GH, d'Azevedo PA, Santos AF, Gales AC, Pignatari AC, Medeiros EA. Intravenous polymyxin B for the treatment of nosocomial pneumonia caused by multidrug-resistant *Pseudomonas aeruginosa*. International Journal of Antimicrobial Agents. 2007;30(4):315-319. DOI: 10.1016/j.ijantimicag.2007.05.017

[60] Kallel H, Hergafi L, Bahloul M. Safety and efficacy of colistin compared with imipenem in the treatment of ventilator-associated pneumonia: a matched case-control study. Intensive Care Medicine. 2007;33(7):1162-1167. DOI: 10.1007/s00134-007-0675-2

[61] Gunderson BW, Ibrahim KH, Hovde LB, Fromm TL, Reed MD, Rotschafer JC. Synergistic activity of colistin and ceftazidime against multiantibiotic-resistant *Pseudomonas aeruginosa* in an in vitro pharmacodynamic model. Antimicrobial Agents and Chemotherapy.2003;47(3):905-909. DOI: 10.1128/aac.47.3.905-909.2003

[62] Zavascki AP, Goldani LZ, Li J, Natio RL. Polymyxin B for the treatment of multidrug-resistant pathogens: a critical review. Antimicrobial Agents and Chemotherapy. 2007 ;60(6):1206-1215. DOI: 10.1093/jac/dkm357

[63] D'Souza BB, Padmaraj SR, Rekha PD, Tellis RC, Prabhu S, Pothen P. In vitro synergistic activity of colistin and ceftazidime or ciprofloxacin against multidrug-resistant clinical strains of *Pseudomonas aeruginosa*. Microbial Drug Resistance. 2014;20(6):550-554. DOI: 10.1089/mdr.2014.0006

[64] Bialvaei AZ, Samadi Kafil H. Colistin, mechanisms and prevalence of resistance. Current Medical Research and Opinion. 2015;31(4):707-721. DOI: 10.1185/03007995.2015.1018989

[65] Falagas ME, Kastoris AC, Karageorgopoulos DE, Rafailidis PI. Fosfomycin for the treatment of infections caused by multidrug-resistant non-fermenting gram-negative bacilli: a systematic review of microbiological, animal and clinical studies. International Journal of Antimicrobial Agents. 2009;34(2):111-120. doi: 10.1016/j.ijantimicag.2009.03.009

[66] Patwardhan V, Singh S. Fosfomycin for the treatment of drug resistant urinary tract infections: potential of an old drug not explored fully. International Urology and Nephrology. 2017;49(9):1637-1643. DOI: 10.1007/s11255-017-1627-6

[67] Okazaki M, Suzuki K, Asano N. Effectiveness of fosfomycin combined with other antimicrobial agents against multidrug-resistant *Pseudomonas aeruginosa* isolates using the efficacy time index assay. Journal of Infection and Chemotherapy . 2002;8(1):37-42. DOI: 10.1007/s101560200004

[68] Lambert PA. Mechanisms of antibiotic resistance in *Pseudomonas aeruginosa*. . **Royal Society of Medicine Press**. 2002;95 (Suppl 41):22-26. PMCID: PMC1308633

[69] Strateva T, Yordanov D. *P. aeruginosa* a phenomenon of bacterial resistance. International Journal of Medical Microbiology. 2009;58(Pt 9):1133-48. DOI: 10.1099/jmm.0.009142-0

[70] Livermore DM. Multiple mechanisms of antimicrobial resistance in *Pseudomonas aeruginosa*: our worst nightmare. Clinical Infectious Diseases. 2002;34(5):634-40. DOI: 10.1086/338782

[71] Masuda N, Ohya S. Cross-resistance to meropenem, cephems, and quinolones in *P. aeruginosa*. Antimicrobial Agents and Chemotherapy. 1992; 36(9):1847-51. DOI: 10.1128/AAC.36.9.1847

[72] Mizuta M, Linkin DR, Nachamkin I. Identification of optimal combinations for empirical dual antimicrobial therapy of *Pseudomonas aeruginosa* infection: potential role of a Combination Antibiogram. Infection Control and Hospital Epidemiology. 2006; 27(4):413-5. DOI: 10.1086/503175

[73] Rahal JJ. Novel antibiotic combinations against infections with almost completely resistant *P. aeruginosa* and *Acinetobacter* species. Clinical Infectious Diseases. 2006;43 Suppl 2:S95-99. DOI: 10.1086/504486

[74] Par SY, Park HJ, Moon SM. Impact of adequate empirical combination therapy on mortality from bacteremic *P. aeruginosa* pneumonia. BMC Infectious Diseases.2012; 12(1):308. DOI: 10.1186/1471-2334-12-308

[75] Das T, Sehar, S, Manefield, M. The roles of extracellular DNA in the structural integrity of extracellular polymeric substance and bacterial biofilm developmen. Environmental Microbiology Reports.2013; 5(6):778-86. DOI: 10.1111/1758-2229.12085

[76] Donlan, R.M., 2002. Biofilms: microbial life on surfaces. Emerging Infectious Diseases 8(9):881-90. DOI: 10.3201/eid0809.020063

[77] Stewart PS, Costerton JW. Antibiotic resistance of bacteria in biofilms. Lancet. 2001;358 (9276):135-138. DOI: 10.1016/S0140-6736(01)05321-1

[78] Stewart PS. Mechanisms of antibiotic resistance in bacterial biofilms. International Journal of Medical Microbiology. 2002; 292(2):107-113. DOI: 10.1078/1438-4221-00196

[79] Anderl JN, Franklin MJ, Stewart PS. Role of antibiotic penetration limitation in *Klebsiella pneumoniae* biofilm resistance to ampicillin and ciprofloxacin. Antimicrobial Agents Chemotherapy. 2000; 44(7): 1818-1824. DOI: 10.1128/aac.44.7.1818-1824.2000

[80] Walters 3rd, M.C., Roe, F., Bugnicourt, A., Franklin, M.J., Stewart, P.S., 2003. Contributions of antibiotic penetration, oxygen limitation, and low metabolic activity to tolerance of *Pseudomonas aeruginosa* biofilms to ciprofloxacin and tobramycin. Antimicrobial Agents Chemotherapy. 2003;47(1):317-23.DOI: 10.1128/aac.47.1.317-323.2003

[81] Rasamiravaka T, Labtani Q, Duez P, El Jaziri M. The formation of biofilms by *Pseudomonas aeruginosa*: a review of the natural and synthetic compounds interfering with control mechanisms. BioMed Research International. 2015, 759348. 5:759348. doi: 10.1155/2015/759348

[82] Kang D, Turner KE, Kirienko NV. PqsA promotes Pyoverdine production via biofilm formation. Pathogens. 2017;7(1):3. DOI: 10.3390/pathogens7010003

[83] Jyot J, Sonawane A, Wu W, Ramphal R. Genetic mechanisms involved in the repression of flagellar assembly by *P. aeruginosa* in human mucus. Molecular Microbiology. 2007; 63(4) :1026-38 DOI: 10.1111/j. 1365-2958. 2006.05573.x

[84] Guidos RJ. Combating antimicrobial resistance: policy recommendations to

save lives. Clinical Infectious Diseases. 2011;52(5)S397–S428. DOI:10.1093/cid/cir153

[85] Ventola CL. The antibiotic resistance crisis: part 1: causes and threats. P&T. 2015;40(4):277-283. PMC4378521

[86] Gould IM, Bal AM. New antibiotic agents in the pipeline and how they can help overcome microbial resistance. Virulence. 2013;4(2):185-191. DOI: 10.4161/viru.22507

[87] Bartlett JG, Gilbert DN, Spellberg B. Seven ways to preserve the miracle of antibiotics. Clinical Infectious Diseases. 2013;56(10):1445-1450. DOI; 10.1093/cid/cit070

[88] Economou V, Gousia P. Agriculture and food animals as a source of antimicrobial-resistant bacteria. Infection Drug Resistance. 2015; 8: 49 61. DOI: 10.2147/IDR.S5577

[89] Maron DF, Smith TJ, Nachman KE. Restrictions on antimicrobial use in food animal production: an international regulatory and economic survey. Global Health. 2013;9:48. 2013 Oct 16;9:48. DOI: 10.1186/1744-8603-9-48

[90] Toke O. Antimicrobial peptides: new candidates in the fight against bacterial infections. Biopolymers. 2005;80(6):717-35. DOI: 10.1002/bip.20286

[91] Park SC, Park Y, Hahm KS. The role of antimicrobial peptides in preventing multidrug-resistant bacterial infections and biofilm formation. International Journal of Molecular Sciences. 2011. 12(9):5971-5992 DOI: 10.3390/ijms12095971

[92] Hancock RE, Haney EF, Gill EE. The immunology of host defence peptides: beyond antimicrobial activity. Nature reviews Immunology. 2016; 16(5):321-34.DOI: 10.1038/nri.2016.29

[93] Papareddy P, Kasetty G, Kalle M, Bhongir RK, Morgelin M. NLF20: an antimicrobial peptide with therapeutic potential against invasive *Pseudomonas aeruginosa* infection. Journal of Antimicrobial Chemotherapy.2016;71(1):170-180, DOI: 10.1093/jac/dkv322

[94] Grassi L, Maisetta G, Esin S, Batoni G. Combination strategies to enhance the efficacy of antimicrobial peptides against bacterial biofilms. Frontiers of Microbiology. 2017; 8: 2409. DOI: 10.3389/fmicb.2017.02409

[95] Hirt H, Gorr SU. Antimicrobial peptide GL13K is effective in reducing biofilms of *P. aeruginosa*. Antimicrobial Agents Chemotherapy. 2013; 57(10): 4903-4910. DOI: 10.1128/AAC.00311-13

[96] Zheng Z, Tharmalingam N, Liu Q, Jayamani E, Kim W. Synergistic efficacy of *Aedes aegypti* antimicrobial peptide Cecropin A2 and Tetracycline against *Pseudomonas aeruginosa*. Antimicrobial Agents Chemotherapy. 2017; 61(7): DOI: 10.1128/AAC.00686-17

[97] Clokie MR, Millard AD, Letarov AV, Heaphy S. Phages in nature. Bacteriophage. 2011;1(1):31-45. DOI: 10.4161/bact.1.1.14942

[98] Kutateladze M, Adamia R. Phage therapy experience at the Eliava Institute. Médecine et Maladies Infectieuses. 2008; 38(8): 426-430. DOI: 10.1016/j.medmal.2008.06.023

[99] Villarroel J, Larsen M, Kilstrup M: Metagenomic Analysis of Therapeutic PYO Phage Cocktails from 1997 to 2014. Viruses. 2017; 9(11): pii: E328. DOI: 10.3390/v9110328

[100] Jennes S, Merabishvili M, Soentjens P. Use of bacteriophages in the treatment of colistin-only-sensitive *Pseudomonas aeruginosa* septicaemia in a patient with acute kidney injury-a case report. Critical Care. 2017; 21(1): 129. DOI: 10. 1186/s13054-017- 1709-y

[101] Ly-Chatain MH. The factors affecting effectiveness of treatment in phages therapy. Frontiers Microbiology. 2014; 5: 51. DOI: 10.3389/fmicb.2014.00051

[102] Pires DP, Vilas Boas D, Sillankorva S, Azeredo J. Phage therapy: a step forward in the treatment of *P. aeruginosa* infections. Journal of virology. 2015;89(15):7449-56. DOI: 10.1128/JVI.00385-15

[103] Morello E, Saussereau E, Maura D, Huerre M, Touqui L, Debarbieux L. Pulmonary bacteriophage therapy on *Pseudomonas aeruginosa* cystic fibrosis strains: first steps towards treatment and prevention. PloS one. 2011;6(2). DOI: 10.1371/journal.pone.0016963

[104] Pires DP, Cleto S, Sillankorva S, Azeredo J, Lu TK. Genetically engineered phages: a review of advances over the last decade. Microbiology and Molecular Biology Reviews. 2016;80(3):523-43. DOI: 10.1128/MMBR.00069-15

[105] Vandenheuvel D, Lavigne R, Brussow H. Bacteriophage therapy: advances in formulation strategies and human clinical trials. Annual Review of Virology. 2015; 2(1):599-618. DOI: 10.1146/annurev-virology-100114-054915

[106] Wright A, Hawkins CH, Anggard EE, Harper DR. A controlled clinical trial of a therapeutic bacteriophage preparation in chronic otitis due to antibiotic-resistant *P aeruginosa*; a preliminary report of efficacy. Clinical Otolaryngology. 2009;34(4):349-57. DOI: 10.1111/j.1749-4486.2009.01973.x

[107] Christiaen SE, Matthijs N, Zhang XH, Nelis HJ, Bossier P, Coenye T. Bacteria that inhibit quorum sensing decrease biofilm formation and virulence in *Pseudomonas aeruginosa* PAO1. Pathogens and Disease. 2014; 70(3)217-279. DOI: 10.1111/2049-632X.12124

[108] Rutherford ST, Bassler BL. Bacterial quorum sensing: its role in virulence and possibilities for its control. Cold Spring Harb Perspective in Medicine. 2012; 2(11): a012427. DOI: 10.1101/cshperspect.a012427

[109] Kalia VC. Quorum sensing inhibitors: an overview. Biotechnology advances . 2013;31(2):224-245. DOI: 10.1016/j.biotechadv.2012.10.004

[110] Gokalsin B, Aksoydan B, Erman B, Sesal NC. Reducing virulence and biofilm of *Pseudomonas aeruginosa* by potential Quorum Sensing Inhibitor Carotenoid: Zeaxanthin. Microbial ecology. 2017 Aug;74(2):466-473.DOI: 10.1007/s00248-017-0949-3

[111] Paczkowski JE, Mukherjee S, McCready AR, Cong JP, Aquino CJ. Flavonoids suppress *Pseudomonas aeruginosa* virulence through Allosteric inhibition of Quorum-sensing receptors. The Journal of biological chemistry. 2017 ;292(10):4064-4076. DOI: 10.1074/jbc.M116.770552

[112] Ma L, Terwilliger A, Maresso AW. Iron and zinc exploitation during bacterial pathogenesis. Metallomics. 2015;7(12):1541-1554. DOI: 10.1039/c5mt00170f

[113] Reid DW, Carroll V, O'May C, Champion A, Kirov SM. Increased airway iron as a potential factor in the persistence of *Pseudomonas aeruginosa* infection in cystic fibrosis. The European Respiratory Journal. 2007;30(2):286-92. DOI: 10.1183/09031936.00154006

[114] Cornelis P, Dingemans J. *Pseudomonas aeruginosa* adapts its iron uptake strategies in function of the type of infections. Frontiers in Cellular and Infection Microbiology. 2013; 3:75. DOI: 10. 3389/fcimb. 2013.00075

[115] O'May CY, Sanderson K, Roddam LF, Kirov SM, Reid DW. Iron-binding compounds impair *Pseudomonas aeruginosa* biofilm formation, especially under anaerobic conditions. Journal of Medical Microbiology. 2009; 58(Pt 6):765-73. DOI: 10.1099/jmm.0.004416-0

[116] Minandri F, Bonchi C, Frangipani E, Imperi F, Visca P. Promises and failures of gallium as an antibacterial agent. Future Microbiology. 2014;9(3):379-97. DOI: 10.2217/fmb.14.3

[117] Jeevanandam J, Barhoum A, Chan YS, Dufresne A, Danquah MK. Review on nanoparticles and nanostructured materials: history, sources, toxicity and regulations. Beilstein Journal of Nanotechnology. 2018;9:1050-1074. DOI: 10.3762/bjnano.9.98. eCollection 2018

[118] Wang L, Hu C, Shao L. The antimicrobial activity of nanoparticles: present situation and prospects for the future. International Journal of Nanomedicine. 2017; 12: 1227-1249. Doi: 10. 2147/IJN.S121956

[119] Salomoni R, Leo P, Montemor AF, Rinaldi BG, Rodrigues M. Antibacterial effect of silver nanoparticles in *P. aeruginosa*. Nanotechnology Science and Applications. 2017;10:115-121. DOI: 10.2147/NSA.S133415

[120] Brown AN, Smith K, Samuels TA, Lu J, Obare SO, Scott ME. Nanoparticles functionalized with ampicillin destroy multiple-antibiotic-resistant isolates of *Pseudomonas aeruginosa* and *Enterobacter aerogenes* and methicillin-resistant *Staphylococcus aureus*. Applied and Environmental Microbiology. 2012; 78(8): 2768-2774. DOI: 10.1128/AEM.06513-11

[121] FAO/WHO. Health and nutritional properties of probiotics in food including powder milk with live lactic acid bacteria. pp. 1-4. Food and Agriculture Organization of the United Nations.

[122] Marco ML, Pavan S, Kleerebezem M. Towards understanding molecular modes of probiotic action. Current Opinion in Biotechnology. 2006;17(2):204-10. DOI: 10.1016/j.copbio.2006.02.005

[123] Ubeda C, Pamer EG. Antibiotics, microbiota, and immune defense. Trends in Immunology. 2012;33(9):459-466. DOI: 10.1016/j.it.2012.05.003

[124] Xu K, Cai H, Shen Y, Ni Q, Chen Y, Hu S, Li L. [Management of corona virus disease-19 (COVID-19): the Zhejiang experience]. Journal of Zhejiang University. Medical Sciences. 2020;49(1):147-157. DOI: 10.3785/j.issn.1008-9292.2020.02.02

[125] Priebe, G.P., Goldberg JB. Vaccines for *Pseudomonas aeruginosa*: a long and winding road. Expert Review of Vaccines. 2014;13(4):507-519. DOI: 10.1586/14760584.2014.890053

[126] Doring G, Pier GB. Vaccines and immunotherapy against *Pseudomonas aeruginosa*. Vaccine. 2008;26(8):1011-1024. DOI: 10.1016/j.vaccine.2007.12.007

[127] Gellatly SL, Hancock RE. *Pseudomonas aeruginosa*: new insights into pathogenesis and host defenses. Pathogens and disease. 2013;67(3):159-73. DOI: 10.1111/2049-632X.12033

[128] Grimwood K., Kyd JM, Owen SJ, Massa HM, Cripps AW. Vaccination against respiratory *P. aeruginosa* infection. Human Vaccines & Immunotherapeutics. 2015;11(1): 14-20. DOI: 10. 4161/hv.34296

[129] Pena C, Suarez C, Tubau F, Dominguez A, Sora M, Pujol M, Gudiol F, Ariza J. Carbapenem-resistant *P. aeruginosa*: factors influencing multidrug-resistant acquisition in

non-critically ill patients. European Journal of Clinical Microbiology & Infectious Diseases. 2009;28(5):519-22. DOI: 10.1007/s10096-008-0645-9

[130] Garnacho-Montero J, Sa-Borges M, Sole-Violan J, Barcenilla F, Escoresca-Ortega A, Ochoa M, Cayuela A, Rello J. Optimal management therapy for *P. aeruginosa* ventilator-associated pneumonia: an observational, multicenter study comparing monotherapy with combination antibiotic therapy. Critical Care Medicine. 2007;35(8):1888-95. DOI: 10. 1097/01. CCM. 0000275389. 31974. 22

[131] Greer, N.D. Doripenem (Doribax): the newest addition to the carbapenems. Proceedings / Baylor University Medical Center. 2008;21(3):337-341. DOI: 10.1080/08998280.2008.11928422

[132] Riera E, Cabot G, Mulet X, Garcia-Castillo M, del Campo R. *P. aeruginosa* carbapenem resistance mechanisms in Spain: impact on the activity of imipenem, meropenem and doripenem. The Journal of Antimicrobial Chemotherapy. 2011;66(9):2022-2027. doi: 10.1093/jac/dkr232

[133] Hilas O, Ezzo DC, Jodlowski TZ. Doripenem (doribax), a new carbapenem antibacterial agent. P&T. 2008. 33(3):134-180. PMCID: PMC2730083

[134] Cox G, Ejim L, Stogios, PJ, Koteva K, Bordeleau E. Plazomicin retains antibiotic activity against most aminoglycoside modifying enzymes. ACS infectious diseases. 2018;4(6):980-987. DOI: 10.1021/acsinfecdis.8b00001

[135] Karaiskos I, Souli M, Giamarellou H. Plazomicin: an investigational therapy for the treatment of urinary tract infections. Expert Opinion on Investigational Drugs. 2015;24(11):1501-11. DOI: 10.1517/13543784.2015.1095180

[136] Srinivas N, Jetter P, Ueberbacher BJ, Werneburg M, Zerbe K. Peptidomimetic antibiotics target outer-membrane biogenesis in *Pseudomonas aeruginosa.* Science. 2010;327(5968):1010-1013. DOI: 10.1126/science.1182749

[137] Zheng Panga, Renee Raudonisb, Bernard R. Glickc, Tong-Jun Lina bd, Zhenyu Cheng. Antibiotic resistance in *Pseudomonas aeruginosa*: mechanisms and alternative therapeutic strategies. Biotechnology Advances. 2019;37(1):177-192. DOI: 10.1016/j.biotechadv.2018.11.013

[138] Michael EC, Margaret RD, Alina MH, Monica CG. Novel Therapeutic Strategies Applied to *P. aeruginosa* Infections in Cystic Fibrosis. Materials 2019, 12, 4093; DOI:10.3390/ma12244093

[139] Zhanel GG, Chung P, Adam H, Zelenitsky S, Denisuik A, Schweizer F, Lagace-Wiens PR, Rubinstein E, Gin AS, Walkty A, Hoban DJ, Lynch JP III, Karlowsky JA. Ceftolozane/tazobactam: a novel cephalosporin/beta-lactamase inhibitor combination with activity against multidrug-resistant gram-negative bacilli. Drugs. 2014;74(1):31-51. DOI: 10.1007/s40265-013-0168-2

[140] Solomkin J, Hershberger E, Miller B, Popejoy M, Friedland I, Steenbergen J, Yoon M, Collins S, Yuan G, Barie PS, Eckmann C. Ceftolozane/tazobactam plus metronidazole for complicated intra-abdominal infections in an era of multidrug resistance: results from a randomized, double-blind, phase 3 trial (ASPECT-cIAI). Clinical Infectious Diseases. 2015; 60(10): 1462-1471. DOI: 10. 1093/cid/civ097

[141] Aktas Z, Kayacan C, Oncul O. In vitro activity of avibactam (NXL104) in combination with beta-lactams against Gram-negative bacteria, including

OXA-48 beta-lactamase-producing *K. pneumoniae*. International Journal of Antimicrobial Agents. 2012;39(1):86-9. DOI: 10.1016/j.ijantimicag.2011.09.012

[142] Torres A, Zhong N, Pachl J, Timsit JF, Kollef M, Chen Z, Song J, Taylor D, Laud PJ, Stone GG, Chow JW. Ceftazidime-avibactam versus meropenem in nosocomial pneumonia, including ventilator-associated pneumonia (REPROVE): a randomised, double-blind, phase 3 non-inferiority trial. The Lancet. Infectious Diseases. 2018;18(3):285-95. DOI: 10.1016/S1473-3099(17)30747-8

[143] Lapuebla A, Abdallah M, Olafisoye O, Cortes C, Urban C, Landman D, Quale J. Activity of imipenem with relebactam against gram-negative pathogens from New York City. Antimicrobial Agents and Chemotherapy. 2015; 59 (8): 5029-31. DOI: 10.1128/AAC. 00830-15

[144] Wright H, Bonomo RA, Paterson DL. New agents for the treatment of infections with Gram-negative bacteria: restoring the miracle or false dawn. Clinical Microbial Infection. 2017;23(10):704-12. DOI: 10.1016/j.cmi.2017.09.001